THE
NELSON
CANADIAN SCHOOL MATHEMATICS DICTIONARY

THE NELSON CANADIAN SCHOOL MATHEMATICS DICTIONARY

JOHN A. FYFIELD AND DUDLEY BLANE

CONSULTANT

SHIRLEY FAIRFIELD
SCARBOROUGH BOARD OF EDUCATION

Nelson Canada

I(T)P An International Thomson Publishing Company

Toronto • Albany • Bonn • Boston • Cincinnati • Detroit • London • Madrid • Melbourne
Mexico City • New York • Pacific Grove • Paris • San Francisco • Singapore • Tokyo • Washington

I(T)P™
International Thomson Publishing
The ITP logo is a trademark under licence

© John A. Fyfield and Dudley Blane, 1995

Published in 1995 by
Nelson Canada
A division of Thomson Canada Limited
1120 Birchmount Road
Scarborough, Ontario M1K 5G4

All rights reserved. It is unlawful to reproduce by any means any portion of this book without the written permission of the publisher.

ISBN 0-17-604800-6

Canadian Cataloguing in Publication Data
Fyfield, John A.
  The Nelson Canadian school mathematics dictionary

ISBN 0-17-604800-6

1. Mathematics - Dictionaries. I. Blane, Dudley.
II. Title.

QA5.F94 1995   510'.3   C95-930642-0

1 2 3 4 5 (WC) 99 98 97 96 95

Contents

Introduction vii

Table of symbols ix

Entries 1

Appendix 1 Table of SI prefixes 248

Appendix 2 Table of other prefixes used in mathematics 249

Appendix 3 Significant mathematicians 251

Introduction

Mathematics is one of the oldest branches of learning in the history of the human race, yet it is still modern and is still developing. More than ever, it is essential for research and development in other fields of study, notably in the sciences and technology; but it is, at the same time, an adventure in its own right. Many of us who become interested in mathematics find that it deepens our understanding of the universe in which we find ourselves.

This dictionary has been compiled with these thoughts in mind; but, being a dictionary (and not an encyclopedia or textbook), it focuses on the language of mathematics and aims to present a set of concise, accurate and consistent definitions suitable for school students. Care has been taken to ensure that, while they may need to be augmented in later years, all definitions and explanations will remain true in the context of more advanced mathematics.

Although mathematics is far more than merely a language, its special way of using words and symbols is one source of its strength. To recognise this claim and to start learning the language is to set your foot on the path to success in the subject.

How to use this book

The dictionary entries are arranged alphabetically and most of them take the following form:

- Head word or phrase.
- Part of speech — (a) if an adjective, (v) if a verb, otherwise, a noun.
- Pronunciation: [...], where it may be helpful (see guide below).
- Brief definition.
- Further description and explanation, often with examples, diagrams, graphs. Useful cross-references to other head words are given in **bold** print. If the head word has several distinct uses, these are numbered.
- Origin of the head word, (its derivation), where this may be helpful to a greater understanding of meaning.

Pronunciation

The approximate pronunciation of some head words is given in square brackets, using the following rules:

- Syllables are separated by hyphens, with the main accented syllable printed in *italics*.

Introduction

- g is always pronounced g as in *g*ot.
- s is pronounced as in *s*et, except that sh is pronounced as in *sh*ot.
- zh is pronounced as s in plea*s*ure.
- th is used for th as in *th*in and for th as in *th*is.
- c,q,x are not used.
- a,e,i,o,u, are as in *a*t, *e*nd, *i*t, *o*f, *u*p, although in some unaccented syllables they may be pronounced as er in fath*er*

    ee as ee in f*ee*t   eer as eer in b*eer*
    ay as ay in m*ay*   air as air in f*air*
    ah as a in f*a*ther   ar as ar in c*ar*t
    ow as ow in h*ow*   er as er in h*er*
    y as y in b*y*   oh as o in r*o*pe
    or as or in f*or*   oy as oy in b*oy*
    oo as oo in b*oo*t   yoo as u in t*u*ne

Table of symbols

Note: The **bold** part of each entry in column 1 is the symbol named in column 2. Column 3 shows how to say the symbol (usually in context). Column 4 gives the main dictionary reference, and this should be consulted for meaning, explanation and cross-references.

1 Symbol	*2* Name	*3* Spoken	*4* Dictionary reference
Numbers and sets			
+3	addition sign plus sign positive sign	'plus 3' or 'positive 3'	**plus sign**
−3	subtraction sign minus sign negative sign	'minus 3' or 'negative 3'	**minus sign**
2**.**3 or 2**·**3 or 2**,**3	decimal point	'2 point 3'	**decimal**
±0.5		'plus or minus point 5'	**error**
π	pi	'pi'	**pi**
e	e	'e'	**e**
i	square root of -1	'i'	**i**
∞	infinity	'infinity'	**infinity**
∅	empty set or null set	'the empty set'	**empty**
U	universal set	'universal set'	**universal set**
|x|, **mod** x	absolute value	'mod x'	**absolute**
$\bar{x}$	bar	'x bar'	**bar**
A′	prime	'A prime'	**complement, prime**
a_n	nth term	'a sub n'	**sequence**
S_n	series sum	'sum to n terms'	**arithmetic series, geometric series**
{x,y**}**	set sign	'set of x and y'	**set**
(...**)**	ordinary brackets		**brackets**
[...**]**	square brackets		**brackets**

Table of symbols

1 Symbol	2 Name	3 Spoken	4 Dictionary reference
$\{...\}$	curly brackets, braces		**brackets**
$\overline{b+c}$	vinculum	'vinculum b plus c'	**vinculum**
Operations			
$a + b$	addition sign	'a plus b'	**addition**
$a - b$	subtraction sign	'a minus b'	**subtraction**
$a \times b$	multiplication sign	'a multiplied by b'	**multiplication**
$a \cdot b$	dot product	'a dot b'	**multiplication**
$\frac{a}{b}, a/b, a \div b$	division sign	'a over b', 'a divided by b'	**division**
$a : b$	ratio sign	'a is to b'	**ratio**
$16:8::10:5$	proportion sign	'16 is to 8 as 10 is to 5'	**proportion**
$\begin{vmatrix} a & b \\ c & d \end{vmatrix}$	matrix sign	'2 by 2 matrix, ab, cd'	**matrix**
Σ	sigma	'the sum of'	**sum**
$\sum_{i=1}^{n} a_i$	summation	'the summation from i equals 1 to n of a sub i'	**sigma**
$S \rightarrow 2$	approaches	'S approaches 2'	**limit**
$\lim_{n \to \infty}$	limit	'in the limit, as n approaches ∞'	**limit**
$x\%$	percentage	'x per cent'	**per cent**
$\int$	integral sign	'integral'	**integral**
Relations			
$a = b$	equal sign	'a equals b', 'a is equal to b'	**equal**
$a \equiv b$	identity, congruence	'a is identical to b', 'a is congruent with b'	**identity** **congruent**
$a \neq b$	inequality sign	'a is not equal to b'	**equal**
$a \approx b, a \doteqdot b$	approximation signs	'a is approximately equal to b'	**approximation**
$a > b$		'a is greater than b'	**inequality**

Table of symbols

1 Symbol	2 Name	3 Spoken	4 Dictionary reference
$a \geq b$		'a is greater than or equal to b'	**inequality**
$a < b$		'a is less than b'	**inequality**
$a \leq b$		'a is less than or equal to b'	**inequality**
$y \propto x$	variation sign	'y varies as x', 'y is proportional to x'	**direct proportion**
$y \propto \frac{1}{x}$	variation sign	'y varies inversely as x'	**inverse proportion**
$P \cup Q$	set union	'P cup Q', 'P union Q'	**union**
$P \cap Q$	set intersection	'P cap Q', 'intersection of P and Q'	**intersection**
$a \in P$	element sign	'a is an element of P'	**element**
$B \subset A$	subset	sign 'B is a subset of A'	**subset**
iff	biconditional	'if and only if'	**iff**

Functions

Symbol	Name	Spoken	Dictionary reference
f(x)	function	'function f of x'	**function**
log x	logarithm	'log x'	**logarithm**
$\log_{10} x$	common logarithm	'log to the base 10 of x'	**logarithm**
$\log_e x$	natural logarithm	'log to the base e of x'	**logarithm**
ln x	natural logarithm	'natural logarithm of x'	**logarithm**
antilog y	antilogarithm	'antilog y'	**antilogarithm**
$\sqrt{a}$	square root	'square root of a', 'root a'	**root**
$\sqrt[3]{a}$	cube root	'cube root of a'	**cube root**
$\sqrt[n]{a}$	n^{th} root	'n^{th} root of a'	**root**
a^2	2nd power, square	'a squared'	**square**
a^3	3rd power, cube	'a cubed'	**cube**
a^n	n^{th} power	'a raised to the power n', 'a to the n'	**power**

Table of symbols

1 Symbol	2 Name	3 Spoken	4 Dictionary reference
$n!$	factorial	'n factorial'	**factorial**
δx or Δx	increment	'delta x'	**increment**
sin x	sine	'sine x'	**sine**
cos x	cosine	'cos x'	**cosine**
cosec x	cosecant	'cosec x'	**cosecant**
sec x	secant	'sec x'	**secant**
tan x	tangent	'tan x'	**tangent**
cot x	cotangent	'cotangent x'	**cotangent**

Geometrical symbols

$\angle AOB$ $A\hat{O}B$	angle	'angle AOB'	**angle**
$AB \perp CD$	perpendicular	'AB is perpendicular to CD'	**perpendicular**
$AB \parallel CD$	parallel	'AB is parallel to CD'	**parallel**
$\vec{F}, \overrightarrow{F}, \mathbf{F}, \overrightarrow{AB}$	vector	'vector F', 'vector AB'	**vector**

Symbols used with units
(see also Table of SI prefixes)

5 m	metre	'5 metres'	**length**
5 g	gram	'5 grams'	**mass**
5 s, 5″	second	'5 seconds'	**second**
5°	degree	'5 degrees'	**angle**
5′	minute	'5 minutes'	**minute**
5 rad	radian	'5 radians'	**radian**
5 sr	steradian	'5 steradians'	**steradian**

Symbols used in logic

p∧q	conjunction	'p and q'	**conjunction**
p∨q	disjunction	'p or q'	**disjunction**

A

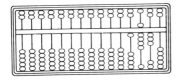

Figure A1

abacus [*ab*-a-kus]   A simple calculating device consisting of beads strung on parallel wires held in a frame. Any given number is represented by a particular arrangement of the beads. With practice, adding is easy; subtraction, multiplication and division are also possible. Figure A1 shows a Chinese abacus.
GREEK   *abax*: a square board

abscissa [ab-*sis*-a]   On a **Cartesian** graph, the distance of a point from the *y*-axis; the *x*-coordinate.
Example
In Figure A2, the abscissa of point P is 4, and the abscissa of Q is -2.
See also   **ordinate**.
LATIN   *abscindere*: to cut off

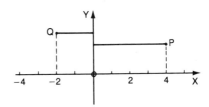

Figure A2

absolute (a) [*ab*-so-loot]   The absolute value of a number is the number without regard for its sign.
Example
$+5$ and -5 both have the absolute value 5. The absolute value of *x* is written |*x*| and is sometimes read as 'mod x'.
See also   **modulus**.

acceleration [ak-sel-er-*ay*-shun] The rate at which the velocity of a moving object changes with respect to time.
Example
A cyclist on a straight road speeds up from 3 metres per second to 4.5 metres per second in 5 seconds. The change in velocity is 1.5 metres per second in 5 seconds, so the average acceleration is 0.3 metre per second per second. This is written as 0.3 m.s^{-2}.
 Acceleration may also be negative, representing a slowing down.
 In more advanced study, an object moving along a curved path is said to be undergoing acceleration, even if its speed remains unchanged.

accuracy [*ak*-yu-ra-see] The accuracy of a number or measurement is an indication of how far an approximate value may differ from a true value.
Example
To say that the square root of 2 is approximately 1.41 means that it is nearer to 1.41 than to either 1.40 or 1.42, or that it lies between 1.405 and 1.415. The degree of accuracy in this case is .005 either way and this is shown by writing 1.41±.005.
See also **approximation**.
LATIN *accuratus*: prepared with care

Achilles and the tortoise [a-*kil*-eez] This is one of the **paradoxes** of the ancient Greek philosopher, **Zeno** (about 500 BC).
 A paradox is a self-contradictory statement. It may at first sight appear logical, while leading to an absurd conclusion.
 Zeno tells the story of the Greek hero, Achilles, who can run 10 times as fast as a tortoise. In a race, he gives the tortoise a 100 metres start, but then cannot win because, in the time it takes him to run the first 100 metres, the

tortoise runs 10 metres. Then, while Achilles runs that 10, the tortoise runs 1 metre more, and so on indefinitely.

This paradox is important in mathematics for the way it introduces us to the idea of a limit.
See also **limit**.

acute [a-*kyoot*]   **1**   An acute angle is smaller than a right angle (90°). **2**   An acute-angled triangle has each one of its angles less than a right angle.
Compare **right-angled triangle, obtuse-angled triangle**.
LATIN   *acutus:* sharp pointed

addend   Any number that is added, e.g. in $3 + 4 + 9$, each of the numbers is an addend.

addition   One of the fundamental operations of mathematics in which two or more numbers are combined to produce their **sum**.

The addition of the numbers a and b produces their sum, $a + b$ (read 'a plus b').

There are special rules for the addition of signed numbers:
- $(-a) + (-b) = -(a + b)$,
  e.g. $(-7) + (-2) = -9$
- $(+a) + (-b) = a - b$,
  e.g. $(+7) + (-2) = 5$
- $(+a) + (-b) = -(b - a)$,
  e.g. $(+3) + (-9) = -6$

In set theory, the comparable operation to addition is the **union** of sets.

addition formulae   Formulae for calculating, in trigonometry, the **sine, cosine, tangent**, etc. of the sum of two angles, given values for the individual angles.
Examples
a $\sin(A + B) = \sin A \cos B + \cos A \sin B$
b $\sin(A - B) = \sin A \cos B - \cos A \sin B$
c $\cos(A + B) = \cos A \cos B - \sin A \sin B$

adjacent

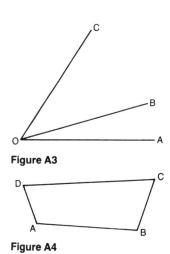

Figure A3

Figure A4

d $\cos(A - B) = \cos A \cos B + \sin A \sin B$

e $\tan(A + B) = \dfrac{\tan A + \tan B}{1 + \tan A \tan B}$

adjacent (a) [a-*jay*-sent]
1 Adjacent angles share a **vertex** and one arm.
Example
Angles AOB and BOC in Figure A3 are adjacent.
2 Adjacent sides of a **polygon** share a vertex.
Example
Sides AB and AD in Figure A4 are adjacent, but AB and DC are not.
LATIN: lying near

algebra The branch of mathematics that uses symbols to study numbers and the relations between them.
    The use of algebraic symbols such as a, b, x, y having variable values makes for greater scope than is possible in arithmetic, which uses only constant numbers such as 5 and $5\frac{1}{2}$.
ARABIAN   *al jebra*: putting parts together

algebraic equation   An equation for which only algebraic operations are needed to solve it.
Example
$3x^2 + 2x - 1 = 0$, but $3x^2 + 1 = \sin x$ is *not* an algebraic equation.

algebraic function   A relation between variables involving only **algebraic operations**.
Example
$y = 3x^2$; $y = ax^2 - bx + c$ are algebraic functions, but $y = \sin x$; $y = 3 \log x$ are *not* algebraic functions.
See also   **function**.

algebraic operations   In ordinary algebra, these are addition, multiplication, subtraction, division, extraction of roots and raising to a power.

The operations of calculating sine, cosine, and logarithms are examples of non-algebraic operations.
See also **transcendental**.

algorithm   A set of steps for finding the solution to a problem. Algorithms are especially important in programming a machine (e.g. a computer) to carry out computations. Sometimes called 'algorism'.
ARABIC   *al-Khwarizmi*: a mathematician of the 9th century

aliquot part [*al*-i-kwot]   A part of a quantity that is an exact **divisor** of the whole.
Examples
a   5 cents is an aliquot part of 1 dollar, because it is exactly 1/20, but 75 cents (i.e. $\frac{3}{4}$ of a dollar) is *not* an aliquot part.
b   The aliquot parts of 12 are 2, 3, 4, 6.
LATIN   *aliquot*: some, several

alternate angles [all-*ter*-nate]   When two straight lines in a plane are cut by a third line, two pairs of alternate angles are formed. In Figure A5, *a* and *b* are a pair of alternate angles. *c* and *d* are another pair.

The special case when $a = b$ and $c = d$ occurs if and only if the lines PQ and RS are parallel.

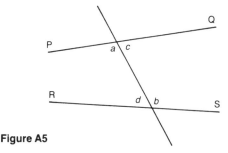

Figure A5

altitude [*al*-ti-tyood]   For a **polygon**, an altitude is the line segment drawn from one

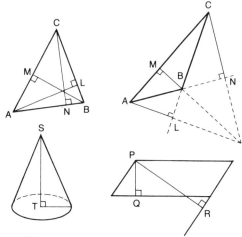

Figure A6

vertex perpendicular to an opposite side. In the two triangle examples shown in Figure A6, the altitudes are AL, BM, CN. An interesting fact is that the three altitudes of any triangle are **concurrent**.

Two altitudes may be drawn from each vertex of a quadrilateral. The example shows two altitudes drawn from P, PQ and PR.

For a solid, the altitude is the line drawn from a vertex to the opposite **face**. ST is the altitude of this cone.

LATIN   *altus*: high

amplitude [*am*-pli-tyood]   The maximum displacement from its mean position of a body undergoing periodic motion. In simple cases,

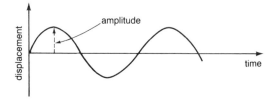

Figure A7

such a motion may be represented graphically by a **sine** curve as shown in Figure A7.

LATIN *amplus*: large

analog or **analogue** (a)   Referring to machines and devices that use physical quantities rather than **digits** for storing and processing information.
Example
A mercury thermometer uses the length of a mercury column to represent the number of degrees of temperature; a clock with hands uses the rotation of its hands to indicate the number of hours and minutes.
See also **digital**.
GREEK *analogia*: proportion

analytical geometry   Another name for **coordinate geometry**.

angle   In plane geometry, an angle is the union of two **rays** (half lines) starting from one point. The two rays are called the sides or arms of the angle, and the common point is called the **vertex** of the angle. The angle in Figure A8 is written ∠AOB.

An angle is a way of showing a change of direction. The size of the angle measures the amount of turning from one direction to another.

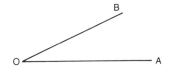
Figure A8

Example
The rotation from direction OA to direction OB in Figure A9 is a rotation through angle AOB.

The most usual measuring system for angles counts a complete rotation as 360 degrees, written 360°.

Angles are classified according to their size as in the following table.

For some purposes, angles may be considered positive if they are measured in an anticlockwise direction and negative in a clockwise direction.

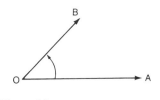
Figure A9

angle of depression

Name	Measure	Illustration
acute angle	between 0° and 90°	
right angle	90°	
obtuse angle	between 90° and 180°	
straight angle	180°	
reflex angle	between 180° and 360°	
perigon	360°	

They may also be unlimited in the number of revolutions they can represent.

Example
Figure A10 may represent any of the three examples shown in Figure A11, and so on.
See also **degree, dihedral, inclination, protractor, radian, steradian.**

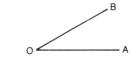

Figure A10

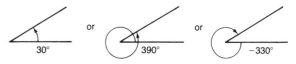

Figure A11

angle of depression — see **elevation**.

angle of elevation — see **elevation**.

angle of inclination — see **inclination**.

annulus [*an*-yoo-lus]   The region between two circles having the same centre.
   If the radii of the circles in Figure A12 are R and r, then the area of the annulus is $\pi R^2 - \pi r^2$ or $\pi(R^2 - r^2)$.
LATIN   a ring

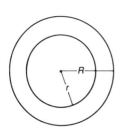

Figure A12

ante meridiem (adverb) [an-tee-mer-*id*-ee-em]   Before midday. Used in quoting time of day, e.g. 9 a.m. means nine o'clock in the morning.
Compare **post meridiem**.
LATIN   *ante*: before, *meridien*: midday

anticlockwise (a)   Turning in a sense opposite to that of the hands of a clock. In some branches of mathematics, an anticlockwise turn is regarded as positive. In Figure A13 the turn from direction OP to direction OQ is an anticlockwise turn. Also known as counterclockwise.

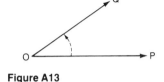

Figure A13

antilogarithm   The inverse of the **logarithm** function. The statement log $x = y$ may be rewritten as antilog $y = x$.
Example
In common logarithms (base = 10), log 100 = 2, antilog 2 = 100.

apex [*ay*-peks]   The point of a figure that is the highest from the base. In the case of a **polygon**, any one **vertex** may be regarded as the apex, depending on which side is to be called the base (see Figure A14).
LATIN   point, summit

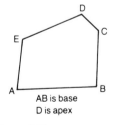 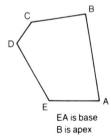

AB is base
D is apex

EA is base
B is apex

Figure A14

approximation   An approximation to a number is another number that is near enough for the purpose. For example, in calculating the area of a circle from its radius, 3.14 may sometimes be used as an approximation to π, while at other times 3.142, depending on the required degree of **accuracy**.

Approximations are also used in measuring. No measurement can ever be regarded as perfectly accurate. When carrying out a series of calculations with approximate numbers or measures, it is very important to avoid too great an error in the result.

The sign for approximation is $\approx$ or $\doteqdot$, e.g. $\sqrt{2} \approx 1.41$, read as 'root two is approximately equal to 1.41'.

See also **significant figures**.

Arabic numerals The symbols 0,1,2,3,4,5,6,7,8,9 used to express numbers in the **decimal** system throughout most of the world today. Their history can be traced back to the Hindu civilization in India over 1200 years ago. Their modern form was developed by Arabian mathematicians through whose publications they were introduced into Europe in the 12th century. **Fibonacci** recommended their use in a book he published in 1202.

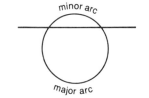

Figure A15

arc Part of a curve. In particular, part of the **circumference** of a circle.

A line drawn through a circle may divide the circumference into two unequal arcs — a major arc and a minor arc, as in Figure A15.

LATIN *arcus*: a bow

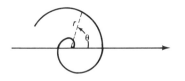

Figure A16

Archimedean spiral [ar-ki-*mee*-di-an *spy*-ral] A spiral drawn according to the rule that its radius increases proportionately to the angle through which it has turned around its starting point.

The equation (see Figure A16) is:
$$r = a\theta \text{ where } a \text{ is constant}$$

Archimedes (287–212 BC) [ar-ki-*mee*-deez] Born and died in Syracuse, Sicily. The greatest scientific and mathematical intellect of the ancient world and one of the greatest of all time. In the words of the mathematician, E. T. Bell: 'Modern mathematics was born in Archimedes and died with him for all of two thousand years. It came to life again in Descartes and Newton.' In mathematics, he laid the foundation for modern calculus and mechanics.

area An amount of surface. It may be plane or curved, but in either case it is two-dimensional.
   The SI unit of area is square metre (m^2). Land area is commonly measured in hectares.
1 hectare (ha) = 10 000 m^2. Some important area formulae:

square: $A = a^2$ rectangle: $A = ab$
circle: $A = \pi r^2$ sphere: $A = 4\pi r^2$

LATIN: a vacant piece of land

arithmetic The branch of mathematics that deals with numerical calculations, such as addition, subtraction, multiplication, division and the extraction of roots.
Compare **algebra**.
LATIN *arithmetica*: the art of counting

arithmetic mean [a-rith-*met*-ik meen]   The arithmetic mean of a collection of n numbers is calculated by adding the numbers together and dividing their sum by n, e.g. arithmetic mean of 1,3,5,7,9 is $\frac{1+3+5+7+9}{5} = \frac{25}{5} = 5.$

   It is known in **statistics** as a measure of **central tendency**. For other such tendencies, *see also* **median**, **mode**. The arithmetic mean is sometimes referred to simply as the average.

arithmetic progression [a-rith-*met*-ik pro-*gresh*-n]   A **sequence** of equally spaced numbers.

Examples
a 1,2,3,4,5,...
b 7,10,13,16,19,...
c 8,5,2,−1,−4,...

In each example, a fixed amount is added to each term in order to generate the next term. This amount is known as the common difference. In **a**, the common difference is 1; in **b** 3; in **c** −3.

In general, an arithmetic progression (AP) can be represented as $a, a + d, a + 2d, ...$, where a is the first term and d is the common difference. The nth term is $a + (n − 1)d$.

arithmetic series [a-rith-*met*-ik *see*-riz] The sum of the terms of an **arithmetic progression**.
Examples
a $1 + 2 + 3 + 4 + 5 + ...$
b $7 + 10 + 13 + 16 + 19 + ...$
c $8 + 5 + 2 + (−1) + (−4) + ...$

The sum of the first five terms (S_5) for each of the above is **a** 15; **b** 65; **c** 10.

In general, the sum of n terms is $S_n = \frac{1}{2}n[2a + (n − 1)d]$, where a is the 1st term and d is the common difference.

array An arrangement of numbers in rows and columns, such as a **matrix**.

ASCII [*as*-kee] These letters stand for *A*merican *S*tandard *C*ode for *I*nformation *I*nterchange, which represents a standard way of assigning numerical codes to numbers, letters and other symbols to be interpreted by a computer.

associative (a) [a-*soh*-si-a-tiv] The property that brackets may be disregarded. Addition and multiplication are associative.
Examples
a $7 + (4 + 3)$ is the same as $(7 + 4) + 3$
b $7 \times (4 \times 3)$ is the same as $(7 \times 4) \times 3$

Subtraction and division are not associative.
Examples
a 7 − (4 − 3) is not the same as (7 − 4) − 3
b 12 ÷ (4 ÷ 2) is not the same as (12 ÷ 4) ÷ 2
The general statement of the associative law is:
$$a * (b * c) = (a * b) * c$$
which is true if * represents addition or multiplication, but is not true if * represents subtraction or division.

assumption A statement to be accepted as true for a particular argument, e.g. in the following discussion, the statement '$x = 3$' is an assumption: Let $x = 3$, then $x^2 + 1 = 10$.
Compare **axiom, premise**.
LATIN *assumere*: to take up

astroid A curve traced out by a point on the circumference of a circle rolling around the inside of another circle of four times its radius (Figure A17). It is an example of a **hypocycloid** with four **cusps**.
Also called **star curve**.
GREEK *astron*: star

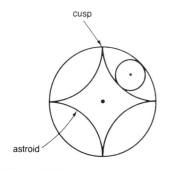

Figure A17

asymmetric (a) [ay-si-*met*-rik] Without **symmetry**, unbalanced.
A shape may have symmetry about one axis but not about another.
Example
The triangle in Figure A18 is asymmetric about the line PQ, but symmetrical about AB.
GREEK *a*: not, *syn*: together, *metron*: measure

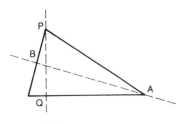

Figure A18

asymptote [*as*-im-tote] A line that is steadily approached but not touched by a curve as its distance from the starting point increases.

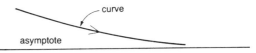

Figure A19

See also **hyperbola**.
GREEK   *a*: not, *syn*: together, *ptotos*: fall

atto-   A prefix meaning one million-million-millionth. Symbol 'a'.
Example
$$1 \text{ attometre} = 10^{-18} \text{ metre}$$
$$1 \text{ am} = 10^{-18} \text{ m}$$
DANISH   *atten*: eighteen

average   Another word for **mean**.
See also   **arithmetic mean, geometric mean, median, mode.** These are all averages, but often it is the first of these that is intended.

axiom   A statement assumed to be true without proof. In logic and mathematics, axioms are the basis from which other statements and theorems are deduced by proof.
Example
One of **Euclid's** axioms is the assertion that 'if equals are added to equals, the results are equal'.

axis [*ak*-sis]   *pl.* axes [*ak*-seez]
1 A line drawn through the centre of a figure. It is usually, but not always, an axis of **symmetry**. The dotted lines in Figure A20 are all axes of symmetry. An axis of symmetry may also be an axis of rotation for a rotating solid.

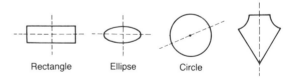

Rectangle   Ellipse   Circle

Figure A20

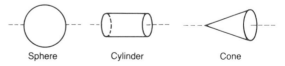

Sphere   Cylinder   Cone

Figure A21

Examples
Figure 21 shows three rotating solids with their axes of symmetry.

2 A line drawn for reference in a **coordinate** system.

Cartesian coordinates in a plane require two axes at right angles to each other, usually called the x-axis (horizontal) and the y-axis (vertical).

In three-dimensional space, three axes are needed, the third one being called the z-axis, which is perpendicular to the x–y plane.

For **polar coordinates** in a plane, only one axis is needed.

B

bar A small horizontal line placed above a symbol. These are some of its uses:
- $\bar{x}$ (*pron.* x bar): the arithmetic mean of a series of values. For example, if x has the values 1,3,4,8, then $\bar{x} = \frac{1}{4}(1 + 3 + 4 + 8) = 4$.
- $\bar{A}$ (*pron.* A bar): vector A.
- $\bar{2}.3$ (*pron.* bar 2 point 3): $-2 + .3$; used in **logarithms** $\bar{2}$ is called a negative characteristic.

bar graph A visual way of showing information in which quantities are represented by bars of equal width.
Example
Figure B1 shows a bar graph, with the different quantities represented by the lengths of the bars.

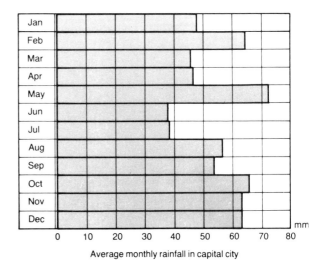

Figure B1

barrel A unit of **capacity**, used mainly for measuring the **volume** of oil. It is approximately 159 litres.

base

1 In a counting system, the base is the number of single digit numerals (including zero) in that system. The base of the **decimal** system is ten, there being ten single digit numerals: 0,1,2,3,4,5,6,7,8,9. The base of the binary system is two, there being only two digits: 0,1. The base of the octal system is eight. In all cases, the base of a system of notation is the number represented by the symbol 10 in that notation.

$10_{decimal}$ = ten 10_{binary} = two 10_{octal} = eight

2 The term is also used in **logarithms**. Because $100 = 10^2$, the logarithm of 100 is said to be 2 with base ten. This is written: $\log_{10} 100 = 2$.

Because $3^5 = 243$, the logarithm of 243 is 5 with base three, written: $\log_3 243 = 5$.

3 In geometry, the base is the bottom line or surface of a figure, as shown in Figure B2.

GREEK   *basis*: stand

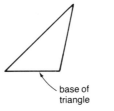
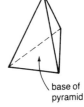

base of triangle   base of pyramid

Figure B2

BASIC A computer programming language. The letters stand for *B*eginner's *A*ll-purpose *S*ymbolic *I*nstruction *C*ode.

bearing The horizontal direction of a line, measured by the angle it makes with the north–south direction.

Example

In Figure B3 the line OP is shown with bearings 30°E, 135°E and 30°W.

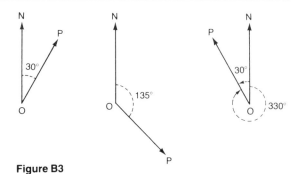

Figure B3

In air navigation, all bearings are measured **clockwise** from north. Thus the bearings in the diagrams would be written as 30°, 135° and 330°.

Magnetic bearings are bearings that are measured from **magnetic north** instead of from true north.

bias [*by*-as] In **statistics**, an unwanted influence on a **sample** that prevents the sample from being truly representative of the **population** from which it is drawn.
FRENCH   *biais*: slant

billion In North American usage, 1 billion = 1 thousand million, i.e., 1 000 000 000 or 10^9. In British usage, 1 billion = 1 million million, i.e. 1 000 000 000 000 or 10^{12}.

In some countries there is uncertainty. France has traditionally interpreted a billion as 10^9, though more recently it has become 10^{12}. Australians now mostly follow American practice (10^9).

bimodal (a) [by-*moh*-dl] In **statistics**, a bimodal distribution shows two distinct peaks of **frequency**.

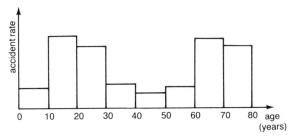

Figure B4

Example
The **histogram** shown in Figure B4 tells that the accident rate peaks for young people and for old people — it is bimodal.

LATIN   *bis*: twice, having two, *modus*: measure

binary notation   Referring to a number system based on the two digits, 0 and 1, and using place value in a similar way to the decimal system. The following table gives a comparison between the decimal system (based on ten digits) and the binary system.

decimal 0 1 2 3 4 5 6 7 8 9 10
binary 0 1 10 11 100 101 110 111 1000 1001 1010 etc.

Binary notation is the basis of digital computing, since 0 and 1 can represent 'off' and 'on' in an electric circuit. In this context, 0 and 1 are referred to as 'bits' (abbreviation for *b*inary dig*it*).
See also **base**.

LATIN   *binarius*: consisting of two

binary operation [*by*-na-ree]   A process applied to a pair of numbers or quantities to produce a single number or quantity.
Example
$3 \times 2 = 6$. Multiplication is a binary operation.
  Addition of numbers and union of sets are other examples of binary operations.
See also **closure**.

binomial [by-*noh*-mi-al] An algebraic expression, which is the sum of two terms.
Example
$3x^2 + 1$ and $5x - 2y$ are binomials.
See also **polynomial, trinomial**.
LATIN *bis*: twice, *nomen*: name

binomial coefficients The numerical factors of the terms in the expansion of $(x + y)^n$, where n is a whole number. For three different values of n, the following table shows this expansion n and the corresponding binomial coefficients.

n		Binomial coefficients
1	$x + y$	1, 1
2	$x^2 + 2xy + y^2$	1, 2, 1
3	$x^3 + 3x^2y + 3xy^2 + y^3$	1, 3, 3, 1

bisect (v) [by-*sekt*] In geometry, to divide into two equal parts. A line cutting an **angle** or a **line segment** into two equal parts is called a bisector of that angle or line segment (see Figure B5).

Figure B5

If a bisector of a line segment cuts at right angles, it is called the **perpendicular bisector** (see Figure B6).
LATIN *bis*: twice, *secare* to cut

Figure B6

bit — **b**inary dig**it** In the binary number system, the two digits 0, 1 are called bits. Strings of bits represent numbers or codes in digital computing.

Example
100101 represents the decimal number 37.
See also **base, binary notation**.

bivariate (a) [by-*vair*-ee-ayt]   In **statistics**, referring to a **distribution** that involves two variables.

box-and-whiskers plot or **boxplot**   In statistics, a diagrammatic way of showing the **spread** of a **distribution**.

On a horizontal number line, a box is drawn with its left and right edges coinciding with the first and third **quartiles** of the distribution. The width of the box thus represents the **interquartile range** of the distribution. The position of the **median** is marked in the box. Furthermore, 'whiskers' are drawn to the left and right of the box, extending as far as the least and greatest value of the distribution variable. The distance between the ends of these whiskers thus represents the **range** of the variable (see Figure B7).

Boxplots are a useful visual method of comparing the characteristics of two or more distributions.

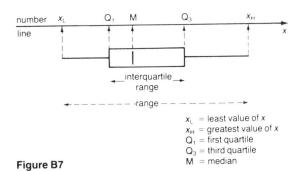

Figure B7

braces   A form of **brackets** used to denote a set in set theory. {a, b, c, d} means the set of the elements a, b, c, d.
LATIN   *brachia*: arms

brackets Brackets are used in mathematics to enclose a number of items intended to be regarded as a single expression. The main types of brackets are:

( )	[ ]	{ }
ordinary brackets	square brackets	curly brackets
(parentheses)		(braces)

If more than one type of bracket is needed, they are usually chosen in the above order.
Example
$2[8 - (5 + 1)] = 2[8 - 6] = 2 \times 2 = 4$

Square brackets have a special use in **matrix** algebra, and braces have a special use in the mathematics of **sets**.
LATIN *brachia*: arms

breadth [*bredth*] The distance from one side of a geometrical figure to the other. Another name for width.

byte A unit of information in computing, usually equivalent to a single character, and stored or processed as an 8-**bit** binary number.
Example
A computer disk with a capacity of 20 MB (20 million bytes) can store 20 million characters (letters, numbers, symbols).

C

calculator A machine for performing mathematical operations.
   Most modern calculators are electronic, are small enough to be held in the hand, are operated by pressing keys and display results in a window. Some calculators can be programmed by the operator to undertake special series of calculations, and some print out the results on paper.
See also **computer**.
   LATIN *calculus*: a small stone

calculus [*kal*-kyu-lus] Any system of calculating, but especially the method founded by **Newton** and by **Leibniz** in the 17th century, which is known as the infinitesimal calculus. It rests on the idea of a limit and includes within its field the study of rates of change and the calculation of lengths, areas and volumes of curved lines, surfaces and solids. It is a powerful branch of mathematics, widely used in science, engineering and economics.
   LATIN: A stone used in calculating

calibrate (v) [*kal*-i-brayt] To determine or check the scale markings on a measuring instrument.
   FRENCH *calibre*: a mould

cancel (v) To remove factors or terms in the process of simplifying a fraction or an algebraic equation.
   1 To simplify a fraction, divide **numerator** and **denominator** by a common factor.

Example

$$\frac{10}{15} = \frac{2 \times 5}{3 \times 5} = \frac{2}{3}$$

2 To simplify an equation, subtract a common **term** from each side or divide each side by a common **factor**.
Examples
a $3a + 2b = 2b + 7$
Subtract $2b$ from each side.
$3a = 7$
b $10x = 120$
Divide each side by 10
$x = 12$
The process is known as cancellation.
LATIN *cancellare*: to crisscross, like a lattice

Cantor, Georg (1845–1918) German mathematician, the founder of modern set theory and noted also for theories associated with the number system and infinity.

cap The symbol, ∩, used for the **intersection** of sets.
Example
The intersection of the two sets, P, Q, in Figure C1 is P ∩ Q (read P cap Q).
See also **cup**.

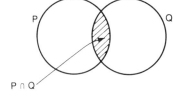

Figure C1

capacity [ka-*pas*-i-tee] The **volume** of liquid or material that can be poured into a container.

Unit of capacity is the litre (L)
1 litre = 1000 millilitres (mL)
1 litre is also equivalent to 1000 cm^3

LATIN *capacitat*: power of holding

carat
1 A unit of **mass**, used mainly for gemstones. It is equal to 200 mg.
2 A measure of gold purity. Pure gold is 24 carat; gold mixed in equal proportion with

another metal is 12 carat; 9 carat gold has 9 parts of gold mixed with 15 parts of another metal.
FRENCH, but originally from GREEK *keras*: horn

cardinal (a) **1** Of chief importance.
Example
The cardinal **compass** points are north, south, east, west.
2 Describing the number of objects in a collection.
Example
The **set** of persons, Tom, Dick and Harriet, has the cardinal number 3.

Two sets have the same cardinal number if and only if they can be put into **one-to-one correspondence** (see Figure C2).
See also **ordinal**.
LATIN *cardinis*: a hinge

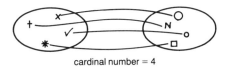

cardinal number = 4

Figure C2

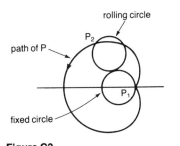

Figure C3

cardioid [*kar*-di-oyd] The path traced out by a point on the **circumference** of one circle that rolls externally without slipping on another circle of the same **radius**. In Figure C3, P1 is the starting point of P, and P2 is another position of P.

The equation is $r = a(1 - \cos \theta)$, where a = radius of each circle, and (r, θ) are the polar coordinates of point P.
GREEK *kardia*: heart

Cartesian (a) [kar-*tee*-zi-an] Derived from **Descartes**, French mathematician.

Cartesian coordinates form a reference system in which any point in a plane is located by its **displacements** from two fixed lines called axes.

The axes are usually at right angles to each other and the common point from which each is measured is called the origin.

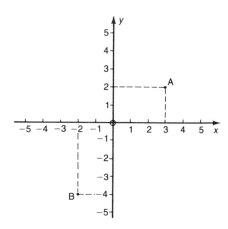

Figure C4

Example
In Figure C4, point A has the location (3,2) and point B has the location (−2,−4). The displacement in the horizontal direction is always stated first and is called the abscissa (or x-coordinate). The displacement in the vertical direction comes second and is called the ordinate (or y-coordinate). The two are known as the coordinates of the point.

For points in space, the system is extended to three dimensions by erecting a third axis at right angles to the other two. Each point then has three coordinates, (x, y, z), as in Figure C5.
See also **coordinate geometry**.

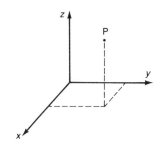

Figure C5

categorical data Information that can be counted under **discrete** headings without being otherwise measurable, e.g. a school list showing countries of origin and the number of students from each country.
Also called **nominal data**.
GREEK *categoria*: accusation

Figure C6

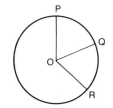

Figure C7

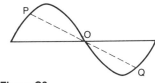

Figure C8

catenary [ka-*tee*-na-ree]   The shape of a uniform heavy flexible cable when it hangs freely between two points, e.g. the shape of power lines strung between pylons.
LATIN   *catena*: a chain

census [*sen*-sus]   A count of all people in a population. Usually, other information such as sex, income, age is gathered at the same time.
LATIN   *censere*: to tax, to estimate

centi-   A prefix meaning one-hundredth. Symbol c.
Example
$$1 \text{ centimetre} = 0.01 \text{ metre}$$
$$1 \text{ cm} = 10^{-2} \text{ m}$$
LATIN   *centum*: hundred

central angle   An **angle** with its vertex at the centre of a circle, e.g. in Figure C7, angles POQ, QOR, POR are all central angles.

central tendency   In describing a statistical distribution, it is often useful to define its central tendency. This means finding some point within the range of collected data about which the data may be thought to be balanced. This point is then regarded as a measure of the central tendency of the distribution. There are three commonly used measures of central tendency: **mean**, **median**, **mode**.
LATIN   *centrum*: centre, *tendere*: to stretch

centre
1   Middle point. For a circle or sphere, this is the point that is an equal distance from every point on the **circumference**.
2   Centre of symmetry. The point around which a curve or figure is **symmetrical**.
Example
In Figure C8, O is the centre of symmetry; $OP = OQ$, etc.

3 Centre of rotation. The point that remains still during the rotation of a figure or object.
4 Centre of gravity. The point about which the weight of an object is evenly distributed. For many calculation purposes, the whole weight of the object may be considered as concentrated at this point.
See also **centroid**.
LATIN *centrum*: centre

centroid Centre of area or volume. If an object is made of the same material throughout, its centroid and its centre of gravity coincide.
 The centroid of a triangle is the point where the **medians** intersect (see Figure C9).
LATIN *centrum*: centre, GREEK *-o-*and-*eides*: like

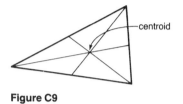

Figure C9

chance The **probability** or likelihood of something happening.
See also **random variable**.
LATIN *cadentia*: falling (of dice)

changing the subject Rearranging a **formula** to make a different **variable** appear on the left-hand side.
Example
In the following formula for finding the number of celsius degrees ($c°$) in a temperature of $f°$ fahrenheit, the subject is c:
$$c = \frac{5}{9}(f - 32)$$
Changing the subject from c to f transforms the formula to:
$$f = \frac{9}{5}c + 32$$

characteristic [ka-rak-ter-*is*-tik] The whole number part of a common **logarithm**.

Example

Number	Logarithm	Characteristic
200	2.3010	2
20	1.3010	1
2	0.3010	0
.2	$\bar{1}.3010$	−1

See also **mantissa**.

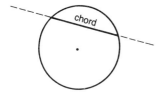

Figure C10

chord [*kord*]   The **line segment** joining two points on a curve. Figure C10 shows a chord of a circle. A circle cannot have a chord larger than its **diameter**.
See also **secant**.
LATIN   *chorda*: a cord

circle   There are two definitions of circle:
1   The set of points in a plane that are all the same distance (r) from a fixed point called the centre. In this case, the circle is a **curve**. Its length is called the **circumference** and is calculated by the formula, $c = 2\pi r$.
2   All the points enclosed by the curve defined above. In this case, the circle is a **region**, and the curve surrounding it is the circumference. The area of the region is calculated by the formula, $A = \pi r^2$.
LATIN   *circus*: a ring

Some parts of a circle

Figure C11

circular argument An attempted proof that relies on evidence that assumes the truth of what is to be proved.
Example
The attempt to prove that the opposite sides of a **parallelogram** are parallel would result in a circular argument, because this property is included in the **definition** of a parallelogram.
Take care! Not all circular arguments are as obvious as this one.

circular functions Another name for the **trigonometric functions**, sine, cosine, tangent, and their **inverses**, cosecant, secant, cotangent. Although the trigonometric functions are often defined using right-angled triangles, they can also be related directly to a circle; hence their description as circular functions.

circumcentre [*ser*-kum-*sen*-ter] The centre of a circle that passes through all the vertices of a **polygon**.
Example
O is the circumcentre of triangle ABC, and P is the circumcentre of pentagon DEFGH, because O and P are the centres of the circles that pass through all the vertices (see Figure C12).

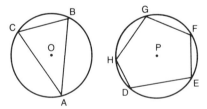

Figure C12

Every triangle has a circumcentre, but not every polygon, e.g. it is not possible to construct one circle that will pass through all the vertices of quadrilateral JKLM in Figure C13, so there is no circumcentre.

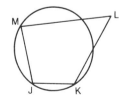

Figure C13

circumscribe

See also **circumcircle, cyclic, incentre**.
LATIN *circum*: around

circumcircle [*ser*-kum-*ser*-kl] A circle that passes through all the vertices of a **polygon**.

All triangles can have a circumcircle, but only some polygons. Those polygons that can have a circumcircle are called **cyclic**. All regular polygons are cyclic.
See also **circumcentre, incircle**.

circumference [ser-*kum*-fer-ens] The boundary of a geometric figure, or the length of the boundary. The term is especially applied to a **circle**.

The length of the circumference (c) of any circle is related to the length of the **diameter** (d) by the formula, $c = \pi d$.
See also **pi**.
LATIN *circum*: around, *ferre*: carry

circumscribe (v) **1** To draw a curved figure (usually a circle) around a **polygon** so that the curve passes through every **vertex** of the polygon.

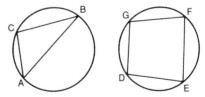

Figure C14

Example
Triangle ABC and quadrilateral DEFG in Figure C14 have each been circumscribed by a circle. It is possible to circumscribe all triangles and all regular polygons by circles, but not all other polygons.
Or

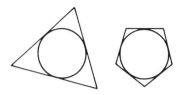

Figure C15

2 To draw a polygon around a curved figure so that every side of the polygon is a **tangent** to the curve.
Example
Figure C15 shows a circle circumscribed by a triangle and a pentagon.
See also **circumcircle, incircle**.
LATIN *circum*: around, *scribere*: write

class interval In **statistics**, a category or division used for grouping a set of observations.
Example
In a particular case, data may range from 0 to 100, with the various observations grouped into class intervals ten units wide, such as *greater than 0 and up to 10, greater than 10 and up to 20*, etc.

Each of these groupings is a class interval, regardless of the number of observations it contains.
See also **histogram**.

clinometer [kly-*nom*-e-ter or klin-*om*-e-ter] An instrument for measuring an angle of **elevation** or depression. It is like a **protractor** with a sighting tube attached.
LATIN *clino-*: slope, and GREEK *metron*: a measure

clockwise (a) Turning in a sense similar to the hands of a clock.
Example
A turn from direction OP to direction OQ in Figure C16 is a clockwise turn. In graphing, this direction is taken as negative.
See also **anticlockwise**.

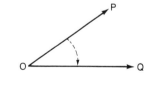

Figure C16

closed
1 A **set** is closed if an operation on any of its elements results in elements that are already within the set.

Example
The set of **natural numbers**, $\{1, 2, 3, \ldots\}$, is closed for the operation of multiplication — if any natural numbers are multiplied together, the result is a natural number. The set of natural numbers is *not* closed for division — if one natural number is divided by another, the result may or may not be a natural number.

2 A closed curve is one that completely surrounds an area.
Example
a **circle** is a closed curve; a **parabola** is not.

coefficient [koh-e-*fish*-ent] A numerical or constant multiplier of the variables in an algebraic term.
Examples
3 is the coefficient in the term $3xy$; $2a$ is the coefficient in the term $2ax$.

More generally, any one **factor** in a product may be regarded as the coefficient of the others.
Examples
In $3xy$, 3 is the coefficient of xy, x is the coefficient of $3y$ and y is the coefficient of $3x$.
See also **binomial coefficient, Pascal's triangle**.

coincident (a) [koh-*in*-se-dent] Occupying the same position.
Example
An **isosceles triangle** (Figure C17) has a **median** and an **altitude** that are coincident. A **scalene** triangle (Figure C18) does not.
LATIN *co-*: together, *incidere*: to happen

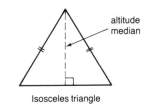

Figure C17 Isosceles triangle

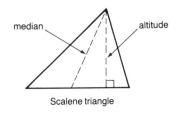

Figure C18 Scalene triangle

co-interior angles A pair of angles formed when two **parallel** lines are crossed by a **transversal**; the co-interior angles are between the parallel lines and on the same side of the transversal.

Co-interior angles are always **supplementary**; that is, they add up to two right angles or 180°.

collinear

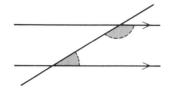

Figure C19

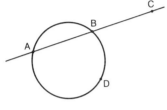

Figure C20

Example
The two angles shaded in Figure C19 form a pair of co-interior angles.

collinear (a) [koh-*lin*-ee-ar]   Lying in the same straight line.
Example
A,B,C in Figure C20 are collinear. A,B,D are not collinear.
LATIN   *co-*: together, *linea*: a line

column [*kol*-um]   A vertical **array** of numbers, as in a *matrix*.
Example
In the matrix shown here, there are two columns: $\begin{matrix} a \\ c \end{matrix}$ and $\begin{matrix} b \\ d \end{matrix}$. $\begin{vmatrix} a & b \\ c & d \end{vmatrix}$
See also   **row**.
LATIN   *columna*: a pillar

column graph   A kind of **graph** in which vertical bars are drawn with different heights to represent different amounts.
Example
The graph in Figure C21 gives information about the monthly rainfall of a city.
See also   **bar graph**.

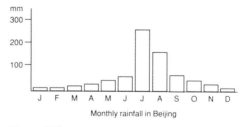

Figure C21

common (a)   An adjective used two ways in mathematics:
1 Shared.
See: **common difference, common factor,**

common multiple, common ratio, common tangent.
2 Ordinary.
See   **common fraction, common logarithm**.
LATIN   *communis*: shared

common difference   The difference between two adjacent terms in an **arithmetic progression** or **series**.

common factor   A number that is a **factor** of each of two or more numbers. This means that it divides into the other numbers without remainder.
Example
5 is a factor of both 10 and 15, so is called a common factor. 10 also has 2 for a factor, but 15 has not, so 2 is not a common factor in this case.
   Common factors occur also in algebra.
Example
$(a + b)$ is a common factor of $(a + b)(a - b)$ and $(a + b)^2$.
See also   **greatest common factor (G C F)**.

common fraction   A fraction expressed as the **ratio** of two whole numbers.
Examples
$\frac{3}{4}, \frac{7}{8}$.
   Common fractions have a **numerator** (top part) and a **denominator** (bottom part):
$\underline{7}$ ← numerator
8 ← denominator
   Common fractions are sometimes called vulgar fractions.
See also   **decimal fraction**.

common logarithm   A **logarithm** using base 10.
Example
$10^3 = 1000$, so 3 is called the common logarithm of 1000. This is written $\log 1000 = 3$, or sometimes $\log_{10} 1000 = 3$.
See also   **natural logarithm**.

common multiple A number that contains every member of a number set a whole number of times.
Example
12 is a common **multiple** of 2, 3, 4. These three numbers have many other common multiples, such as 24 and 60.
 Common multiples occur also in algebra.
Example
$x(x^2 - 1)$ is a common multiple of x and $x + 1$.
See also **least common multiple**.

common ratio For a **geometric progression** or **series**, the number by which any term is multiplied to produce the next term.
Example

Geometric progression	Common ratio
1, 5, 25, 125, ...	5
10, 5, 2.5, 1.25, ...	0.5
2, −4, 8, −16, ...	−2

common tangent A straight line that touches more than one curve.
Example
The straight line in Figure C22 is a common **tangent** to the curves.
 Two circles may have 4, 3, 2, 1 or 0 common tangents as shown in Figure C23.

Figure C22

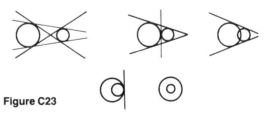

Figure C23

commutative (a) [kom-*yoo*-ta-tiv or *kom*-yoo-tay-tiv] The property that, for certain sequences of operations, order does not matter. Addition and multiplication are commutative:
Examples
a 12 + 3 gives the same result as 3 + 12

b 12 × 3 gives the same result as 3 × 12
Subtraction and division are not commutative:
Examples
a 12 − 3 = 9, but 3 − 12 = −9
b 12 ÷ 3 = 4, but 3 ÷ 12 = $\frac{1}{4}$

The general statement of the commutative property is $a * b = b * a$, which is true if * represents addition or multiplication, but is not true if * represents subtraction or division.

LATIN *commutare*: to exchange

compass [*kum*-pas] An instrument for finding direction. It usually consists of a pivoted magnetic needle, which points towards magnetic north.

LATIN *com-*: together, *passus*: step

compasses (pair of) An instrument used in many geometric constructions, such as:
- drawing circles and arcs
- constructing angles (30°, 60°, 90°, etc.)
- bisecting lines and angles
- transferring lengths and angles
- constructing plane figures.

In traditional geometry, many constructions are carried out on paper using only a pair of compasses and a straight edge.

According to Plato (427?–347?BC) and his followers, these were the only constructions permitted in geometry, so it was not possible for them to trisect an angle or construct a square equal in area to a circle.

complement
1 One angle is called the complement of another angle if the two angles add up to a right angle (see Figure C24).
Example
30° is the complement of 60°; 60° is the complement of 30°.
See also **supplement**.

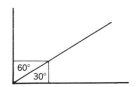
Figure C24

2 A term used in **set** theory. If set A has a

subset B, then any elements of A that are not in this subset make up another subset, called the complement of B. The complement of B is written B'.

Example
The set of the first nine natural numbers $\{1,2,3,4,5,6,7,8,9\}$ has a subset $\{3,6,9\}$. The complement of this subset is $\{1,2,4,5,7,8\}$.

LATIN *complere*: to fill up

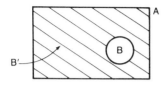

Figure C25

complementary events — see **events**.

completing the square In algebra, the process of adding a constant to a given **quadratic** expression to form a **perfect** square, e.g. $x^2 + 6x + 2$ is not a perfect square, but if 7 is added it becomes $x^2 + 6x + 9$, which is the square of $x + 3$.

In general, if the first two terms are $x^2 + bx$, then the square is completed by adjusting the third term to be $(\frac{1}{2}b)^2$.

The process is useful in constructing graphs of quadratic functions and in solving quadratic equations.

complex number In advanced mathematics, the number system is extended to include numbers that cannot be positioned on the ordinary **number line**. Such numbers are known as complex numbers.

A complex number has a 'real' part and an 'imaginary' part added together in the form $a + ib$. a and b are real numbers and i is interpreted as $\sqrt{-1}$. a is the 'real' part of the complex number and ib is the 'imaginary' part.
See also **real number**.

LATIN *com-*: together, *plectere*: to twine

component [kom-*poh*-nent] Two or more **vectors** acting together, that can effectively replace a given vector are called components of that vector.

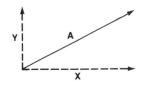

Figure C26

Example
Vector **A** can be replaced by component vectors, **X** and **Y** in figure C26.
See also **resolution**.
LATIN *componere*: to put together

composite graph A diagram showing two or more **bar graphs** combined, as in Figure C27.

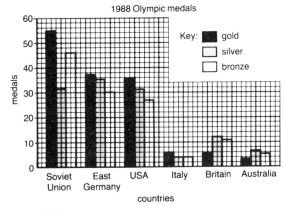

Figure C27

composite number [*kom*-po-zit] Any **integer** that is the result of multiplying two or more other integers together (not counting +1 or −1), e.g. 87 = 29 × 3 is a composite number. A composite number cannot be a **prime** number.

Figure C28

composite shape A geometric shape made up from simpler geometric shapes, e.g. the shape shown in Figure C28 is made from a rectangle and a semicircle.

compound event In statistics, a single **event** that may be considered as being made up of two or more simple parts, e.g. if two coins are tossed, the resulting event will be one of: HH, HT, TH, TT. Each of these four is a compound event.

compound interest **Interest** calculated at regular intervals on a sum of money to which previous interest calculations have been added.
Example
Compound interest calculated yearly on $100 at 10%:
    1st year $100 × 10/100 = $10
    2nd year $110 × 10/100 = $11
    3rd year $121 × 10/100 = $12.10
See also **simple interest**.

computer An electronic device capable of carrying out many tasks that can be expressed as a sequence of precise instructions. These instructions are embodied in a program written in one of a number of computer languages. Each computer language can be represented symbolically in numerical form, so that the working of a computer can be said to be based on mathematical logic.

    A major forerunner to the computer was an 'analytical engine' designed by Charles Babbage (1792-1871) in England. It used punched cards for variables and operators. Augusta Byron (Lady Lovelace) (1815-1852) produced a set of instructions for this machine to calculate a sequence of numbers according to a formula. She may therefore be thought of as the first computer programmer.
    LATIN   *com-*: together, *putare*: to think

concave

Figure C29

concave The shape of a curve or curved surface seen from the inside; hollow (see Figure C29). The reverse of **convex**.
See also **point of inflexion, tangent**.

concentric [kon-*sen*-trik] Circles with the same centre are concentric. Spheres also are concentric if they share the same centre (Figure C30).

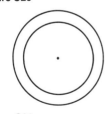

Figure C30

concurrent Sharing a point, as in Figure C31, where all the lines have the point P in common.

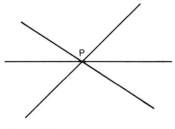

Figure C31

It is interesting that several sets of lines associated with a **triangle** are concurrent: the **medians**, the **altitudes**, the angle **bisectors**, and the perpendicular bisectors of the sides (see Figure C32).

LATIN   *con*: together, *currere*: to run

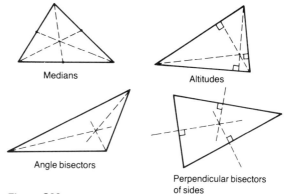

Medians

Altitudes

Angle bisectors

Perpendicular bisectors of sides

Figure C32

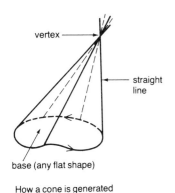

How a cone is generated

Figure C33

cone   A cone is a solid figure whose surface is in two parts: the base is plane (flat), the rest of the surface is curved. The base may be any shape as long as it is plane (Figure C33). To form the curved surface, a straight line passing through a fixed point (the vertex) outside the base traces around the perimeter of the base.

If the base is a circle (Figure C34), the solid is called a circular cone, and, if the line drawn

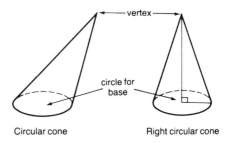

Circular cone

Right circular cone

Figure C34

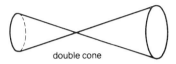

Figure C35

from the centre of the circular base to the vertex is perpendicular to the base, the solid is called a right circular cone.

A right circular cone may be generated by spinning a right-angled triangle, using one of its shorter sides as axis.

Sometimes it is useful to study the double cone formed by extending in opposite senses from the vertex as in Figure C35.

confidence interval In statistics, the interval within which the true value of a **random variable** is estimated to lie with a stated degree of probability.

Example

The mean quoted in the example given under **standard error** lies in the confidence interval 58.5 to 61.5, with a probability of 68%, and in the confidence interval of 57 to 63 with 95% probability.

LATIN *confidere*: to trust

congruent (a) [*kon*-groo-ent]

1 Congruent figures are identical in shape and size. Triangles ABC and DEF in Figure C36 are congruent. Their corresponding sides and angles are equal and one triangle can be made to coincide exactly with the other. Figures with different sizes, even if they have the same shape, are not congruent. The sign for 'is congruent to' is ≅. For example △ABC ≅ △DEF.

2 Congruent numbers share the same

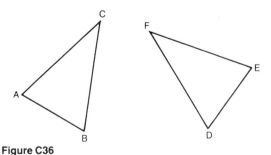

Figure C36

remainder when divided by another number called the **modulus**.

Example
17 and 32 are congruent, modulus 5, because $17 \div 5$ and $32 \div 5$ both give the same remainder (2). This fact is written $17 \cong 32 \pmod 5$.
See also **modular arithmetic**.
LATIN *congruere*: to agree

conic section A curve formed when a right circular **cone** is cut by a plane. The curve will be a **circle**, an **ellipse**, a **parabola** or a **hyperbola**, e.g. if the plane is parallel to the base of the cone, a circle is formed.

The geometry of conic sections was first developed by Greek mathematician, Apollonius (260?–200?BC), who thus laid the foundation for Newton's work in understanding planetary motions 1900 years later.
LATIN *secare*: to cut

conjecture A general conclusion drawn from a number of individual facts. The result of reasoning by **induction**.
LATIN *con-*: together, *jacere*: to throw

conjunction
1 In logic, a **connective**, the equivalent of 'and' that combines two **propositions**. Its symbol is $\wedge$.
2 In logic, the combining of two propositions with 'and'. In this sense, $p \wedge q$ is a conjunction formed from two propositions, p and q; e.g. if p is the proposition 'I am in London' and if q is the proposition 'Today is Thursday' then $p \wedge q$ represents the proposition 'I am in London and today is Thursday'.
See also **disjunction, truth table**.
LATIN *con-*: with, *jungere*: to join

connective In logic, a word or symbol used to join two **propositions**, e.g. the symbols $\wedge$ and $\vee$ are connectives that correspond to the words

'and' and 'or' respectively.
See also **conjunction, disjunction**.

consecutive (a) [kon-*sek*-yoo-tiv]   Following one after the other in regular order.
Examples
a   5,6,7,8 are consecutive whole numbers.
b   2,4,6,8 are consecutive even numbers.
c   3,6,12,24 are consecutive terms of a geometric progression.
LATIN   *consequi*: to follow

constant   **1**   A numerical part of an algebraic expression, e.g. in the expression, $2x + 3$, 2 and 3 are constants.   **2**   A quantity regarded as fixed for a particular calculation, e.g. in the **equation of a line**, $y = mx + c$, m and c are constants for a particular line, while x and y are **variables**.
3   A quantity that is fixed by definition in all circumstances, e.g. π is the ratio of the **circumference** to the **diameter** of all circles.
4   A quantity fixed by the laws of nature, e.g. c, the speed of light.
LATIN   *constare*: to stand firm

contingency table [kon-*tin*-jen-see]   A compact way of presenting the **frequency distribution** of a two-way classification in **statistics**.
Example
If a class of 25 students is classified according to whether they are girl or boy and whether they like or dislike mathematics, the results may be presented as in the contingency table below:

	Boy	Girl	
like	7	10	17
dislike	5	3	8
	12	13	25

LATIN   *contingere*: to take hold of

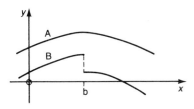

Figure C37

continuous (a) Without a break.
Example
Curve A in Figure C37 represents a continuous function; curve B represents a function that is discontinuous at $x = b$.

Continuous data in **statistics** refer to measures of quantities that allow of continuous change, such as speed and volume, not quantities like the number of pages in a book or students in a class.
See also **discontinuity, discrete**.
LATIN *continere*: to hold together

continuum [kon-*tin*-yoo-um] (*pl.* **continua**) Something whole and connected. The set of real numbers is a continuum, emphasising that it can be represented by a continuous line without gaps.
LATIN *continere*: to hold together

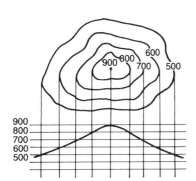

Figure C38

contour A curve drawn through points of equal value.
Example
A curve drawn on a map through points representing places with the same height above sea-level.

Figure C38 shows part of the contour map of a mountain 900 m high. It has contour intervals of 100 m.
FRENCH: outline

contrapositive In logic, a statement formed from another by reversing its order and changing positive to negative and negative to positive.
Example
The contrapositive of 'A triangle that is right-angled is not an equilateral triangle' is 'A triangle that is equilateral is not a right-angled triangle'.

A statement and its contrapositive are logically equivalent, so that the contrapositive of a true statement is itself always true.

convention An agreed rule or practice.
Example
It is a convention that $y \times y$ is written y^2. It is a convention that the horizontal axis of a **Cartesian** graph is called the x-axis and that it is scaled from left to right.

Conventions are decided for the sake of convenience; they are not required by logic.

LATIN *conventio*: a meeting

convergent (a) [kon-*ver*-jent]
1 (geometry) — Aiming towards a point.
Example
After reflection by the concave mirror in Figure C39, the parallel rays are convergent towards point P.

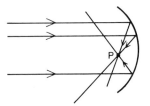

Figure C39

2 (algebra) — Describing a series whose sum has a definite limit, no matter the number of terms.
Example
$1 + \frac{1}{2} + \frac{1}{4} + \frac{1}{8} + \ldots$ does not exceed the value 2 no matter how many terms are added. It is a convergent series. Contrast this with $1 + 2 + 4 + 8 + \ldots$, which is **divergent**.

LATIN *convergere*: to turn towards one another

converse [*kon*-vers] A statement formed from another statement by interchanging subject and predicate. The converse of 'All men are liars' is 'All liars are men'. The converse of a true statement may be true or false.

coordinates

Examples

True statement	Converse	True or false
Squares are always figures with four right angles	Figures with four right angles are always squares	False
Every triangle with three equal sides has three equal angles	Every triangle with three equal angles has three equal sides	True

To argue from a true statement to its converse is not a **valid** form of argument.

LATIN   *convertere*: to turn about

convex (a)   The shape of a curve or curved surface seen from the outside. The reverse of **concave**.

A convex polygon has no interior angle greater than 180°, and any straight line drawn through a convex polygon cuts it in two points only (see Figure C40).

See also   **point of inflexion, tangent**.

LATIN   *convexus*: arched

A convex hexagon   A concave hexagon

Figure C40

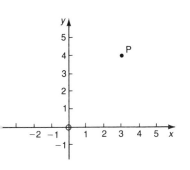

Figure C41

coordinates [koh-*ord*-i-nayts]   A **set** of numbers used to define a position with respect to a **frame of reference**.

Example

The position of point P on the graph shown in Figure C41 is defined as 3 units from the *y*-axis and 4 units from the x-axis. The x-axis and the y-axis make up the frame of reference and the coordinates of P are written as the **ordered pair** (3,4).

A set of three numbers is needed to define a position in three-dimensional space.

See also **abscissa, cartesian, ordinate, polar coordinates**.

LATIN *co-ordinare*: to set in order

coordinate geometry The branch of mathematics in which algebra is applied to geometrical points, lines and figures, the positions of which are defined by **coordinates**.
Example
The line AB in Figure C42 is identified by the algebraic equation, $y = \frac{1}{2}x + 2$.
Also known as analytical geometry.

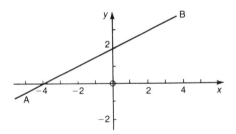

Figure C42

coplanar (a) [koh-*play*-nar] Lying in the same plane.
Example
a All the sides of a **polygon** are coplanar.
b Any two **parallel** lines are coplanar.
c Any two intersecting lines are coplanar.
If lines are neither parallel nor intersecting, they are not coplanar. Instead, they are called **skew**.

corollary [ko-*rol*-a-ree] An additional statement that follows simply from something already proved.
LATIN *corollarium*: a gift

correct to *n* significant figures Accurately representing the first *n* digits of a number.
Example
Because $\sqrt{5} = 2.23606\ldots$, the approximation 2.2361 is correct to five significant figures, and the approximation 2.236 is correct to four significant figures.

correlation In **statistics**, the apparent relation between two sets of data.
Example
In a school, older students are mostly heavier than younger students: there is a correlation between the weights and ages of the students.

Correlation may be strong (high) or weak (low), and positive or negative. Statisticians express the degree of correlation between two sets of related variables as a correlation coefficient, the value of which lies between -1 and $+1$. The value $+1$ represents perfect positive correlation, -1 represents perfect negative correlation, and 0 represents no correlation at all.

If two sets of data show correlation, it is not necessarily true that they are related through cause and effect.
See also **scattergram**.

correspondence A pairing of members of one **set** with members of another by means of a rule.
Example
A set of boys may be paired with a set of girls according to the rule, 'she is his sister'. The correspondence may be illustrated as in Figure C43, showing that Joe has two sisters in the girls' set and Jason has one. The others are not related in this way.

A special case of correspondence is **one-to-one**. An example of this is the correspondence between members of an audience and the seats in a full theatre (see Figure C44). The idea of

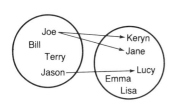

Figure C43

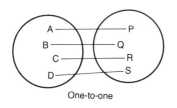

Figure C44

one-to-one correspondence is basic to the notion of **cardinal number**.

LATIN   *correspondere*: answer together

corresponding (a)   Standing in a similar relation to.
Example
In the case of the similar triangles ABC and DEF, (Figure C45) A and D, B and E, C and F are corresponding vertices and the angles at these vertices are corresponding. AB and DE, BC and EF, CA and FD are corresponding sides.

In the case of two lines cut by a third line, four pairs of corresponding angles are formed: *a,e; b,f; d,h; c,g* in Figure C46. If the two lines are parallel, the corresponding angles are equal: $a = e, b = f, d = h, c = g$.

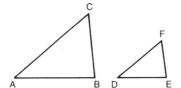

Figure C45

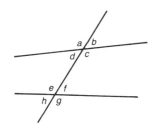

Figure C46

cos [*koz*]   Abbreviation and symbol for **cosine**.

cosecant [koh-*sek*-ant], abbreviation cosec
The **reciprocal** of **sine**. In Figure C47:

$$\operatorname{cosec} \angle NOP = \frac{OP}{PN}$$

The cosecant of a given angle is equal to the **secant** of the **complementary** angle.

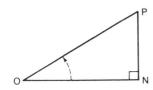

Figure C47

cosine [*koh*-syn], abbreviation **cos**   One of the **circular** (or **trigonometric**) functions. The cosine of the angle θ in the circle shown in Figure C48 is the ratio of the projection of OP on the *x*-axis to the radius OP; i.e. $\cos \theta = x/r$. This value is positive when P is to the right of the *y*-axis and negative when to the left. As P moves around the circle in either direction, θ takes on any value from $-\infty$ to $+\infty$ (see Figure C49). Cos θ will fluctuate in value between -1 and $+1$. It is called a periodic function.

In a right-angled triangle, the cosine of one of the acute angles is the ratio of the adjacent short

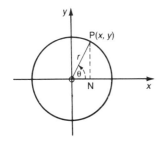

Figure C48

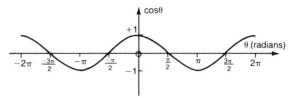

Figure C49   Graph of cos θ

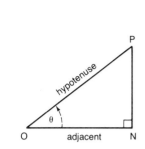

Figure C50

side to the hypotenuse (see Figure C50):
$$\cos \angle \theta = ON/OP$$
The cosine of an angle is the **sine** of its **complementary** angle.

LATIN   *co-*: with, *sinus*: curve

cosine rule   A formula relating the lengths of the three sides and the magnitude of an angle in any triangle.

If a, b, c are the side lengths and A is the magnitude of the angle opposite side a, then
$$a^2 = b^2 + c^2 - 2bc \cos A.$$
Similarly
$$b^2 = c^2 + a^2 - 2ca \cos B.$$
$$c^2 = a^2 + b^2 - 2ab \cos C.$$
Note that if one of the angles (say A) is a right angle, then $\cos A = 0$ and $a^2 + b^2 = c^2$, which is the theorem of **Pythagoras**.
See also   **sine rule**.

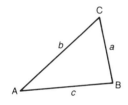

Figure C51

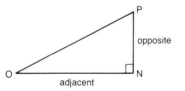

Figure C52

cotangent [koh-*tan*-jent]   The **reciprocal** of the **tangent** of an angle. In Figure C52:
$$\cot \angle NOP = ON/PN$$
The cotangent of an angle is equal to the **tangent** of the **complementary** angle.

LATIN   *co-*: with, *tangere*: to touch

count (v)   **1**   To recite the numerals 1,2,3, ... in order.   **2**   To recite certain **multiples**, e.g. to recite 3,6,9, ... is to count in threes.
3   To associate, by **one-to-one** correspondence, the objects in a **set** with the numerals beginning with 1 and taken in order. This amounts to finding the **cardinal** number of the set:

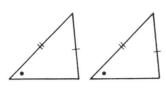
Supporting example

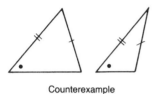
Counterexample

Figure C53

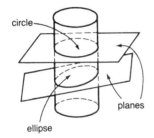
Figure C54

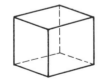
Figure C55

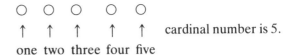
cardinal number is 5.

LATIN *computare*: to reckon

counterexample An example that shows a general statement to be false.
Example
Although it is possible to find examples that seem to support the following statement, it is only necessary to find one counterexample to show that, as a general statement, it is false:
Statement: 'If two sides and an angle of one triangle are equal to two sides and an angle of another triangle, then the two triangles are **congruent**.' Look at Figure C53

cross-section The shape produced when a **plane** cut is made through a solid.
Example
A cross-section of a **right circular cylinder** is either a circle or an ellipse (see Figure C54).

cube 1 In geometry, a regular solid with six square faces (Figure C55). 2 In arithmetic and algebra, the third **power** of a number or expression.
Example
a The cube of 5 is $5 \times 5 \times 5 = 5^3$, read '5 cubed'.
b The cube of n is $n \times n \times n = n^3$, read '$n$ cubed'.
c The cube of $(n-1)$ is written $(n-1)^3$.
Every number has a cube, e.g. the cube of 2.5 is 15.625, and a number that is the cube of a whole number is called a perfect cube. The following are all perfect cubes: 1, 8, 27, 64, 125 ...
Numbers in between these are not perfect cubes.
GREEK *kubos* a cube or a die

cube root The cube root of a number, n, is the number whose third **power** equals n. If $x^3 = n$, then the cube root of n is x. This is written $\sqrt[3]{n} = x$, e.g. the cube root of 125 is 5, because $5 \times 5 \times 5 = 125$. The cube roots of many whole numbers are **irrational** numbers (**surds**), e.g. the cube root of 2 is 1.2599... and the cube root of 10 is 2.15443... Many calculators have a key for the quick calculation of a cube root.

cubic (a) **1** (geometry) — Cube-shaped.
2 (algebra) — Describing a third-degree expression or equation, i.e. one containing a variable raised to the third **power** and none higher; e.g. $2x^3 + 3x - 1$ is a cubic expression and $x^3 + 2x - 1 = 0$ is a cubic equation.
3 Used in naming some three-dimensional measures, e.g. 1 cubic metre (1 m^3) is the volume occupied by a cube having edges of 1 metre.

cubical contents Another name for **capacity**.

cuboid A solid with six rectangular faces, adjacent faces being perpendicular to each other. Box-shaped.
Def. A rectangular **prism**.
GREEK *kubos*: cube, *o-eides*: like

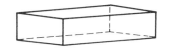

Figure C56

cumulative frequency In a **frequency distribution**, the total of all frequencies up to a certain value of the variable.
Example
The table shows the heights (h cm) of 100 children grouped in 10 cm intervals. The cumulative frequency associated with the range 110–120 is 55, meaning that 55 children have heights in the range up to and including 120 cm.
LATIN *cumulare*: to heap up

h	f	cf
90–100	5	5
100–110	20	25
110–120	30	55
120–130	25	80
130–140	15	95
140–150	5	100

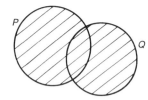

Figure C57

cup   The symbol, ∪, used for the **union** of **sets**.
Example
The union of the two sets, P, Q, in Figure C57 is $P \cup Q$ (read P cup Q).
See also   **cap**.

curve   A path that is nowhere straight. A portion of a curve is called an **arc**. (*Compare*: a portion of a straight line is called a line segment.)

Some interesting and important curves are described in this dictionary. *See*: **astroid, cardioid, catenary, circle, cycloid, ellipse, epicycloid, helix, hyperbola, hypocycloid, Lissajous figures, normal curve, ogive, parabola, sinusoidal.**
LATIN   *curvus*: bent

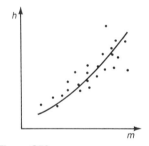

Figure C58

curve of good fit   A curve drawn on a **scattergram** to show in simplified and approximate form the relation between two variables.
Example
The exam scores in history (h) and mathematics (m) are plotted for each student in a class of 25. The curve of best fit smooths out the apparent irregularities (see Figure C58).

curvilinear (a) [ker-vi-*lin*-i-ar]   Formed with curves.
Example
The attempt to draw a triangle on a curved surface such as a sphere produces a figure with curved sides. This is called a curvilinear triangle.

cusp   A point at which two branches of a curve meet, e.g. an **astroid** has four cusps (see Figure C59).
LATIN   *cuspis*: a point

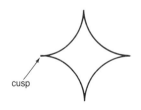

Figure C59

cyclic quadrilateral [*sy*-klik]   A quadrilateral, through the four vertices of which

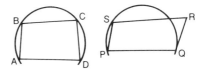

Figure C60

it is possible to draw a circle.
Example
Quadrilateral ABCD in Figure C60 is cyclic; quadrilateral PQRS is not.

The opposite angles of a cyclic quadrilateral add up to two right angles (they are supplementary). This is not true of any other quadrilateral.

GREEK   *kuklos*: a circle

cycloid [*sy*-kloyd]   The curve traced out by a point on the circumference of a circle rolling along a line (see Figure C61).
See also   **epicycloid, hypocycloid**.
GREEK   *kuklos*: a circle, *o-eides*: like

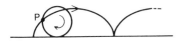

Figure C61

cylinder [*sil*-in-der]   A solid figure whose surface is in three parts: the base and the top are **plane** (flat), parallel and of the same size and shape; the rest of the surface is curved.

The figure is formed between two parallel planes by a line moving around a closed curve and at a fixed angle to the planes. Usually the base and top are circles and the moving line is at right angles to them. The cylinder is then a right circular cylinder (see Figure C62).
GREEK   *kulindros*: a roller

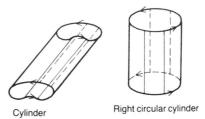

Cylinder   Right circular cylinder

Figure C62

D

data [*day*-ta or *dah*-ta]   Information providing the basis of a discussion from which conclusions may be drawn. Data often take the form of a collection of numbers that can be displayed graphically or in a table. Strictly, the word 'data' is plural, although it is often used collectively as a singular noun.
LATIN   *datum*: given

databank or **database**   A large amount of information stored in an organised way that makes it easy to retrieve individual items. The term is used especially in connection with storage of data in a computer.

deca-   Prefix meaning ten.
See also **deka-**
GREEK   *deka*: ten

decagon [*dek*-a-gon]   A ten-sided polygon. Each interior angle of a **regular** decagon measures 144°.
GREEK   *deka*: ten, *gonia*: angle

decahedron [dek-a-*hee*-dron]   A solid figure with ten plane faces. There is no **regular** decahedron.
GREEK   *deka*: ten, *hedra*: base

deci-   Prefix meaning one-tenth. Symbol d.
Example
$$1 \text{ decimetre} = 0.01 \text{ metre (dm)}$$
$$1 \text{ dm} = 10^{-2} \text{ m}$$
LATIN   *decem*: ten

decimal (a)   Based on 10.

The decimal system is a number system based on 10 and multiples and submultiples of 10: numbers are expressed using just ten **digits**; 0,1,2,3,4,5,6,7,8,9, their value depending on their place in the line, e.g. in the number 347, 3 represents three hundred, 4 represents forty, and 7 represents seven.

The decimal system also extends to numbers less than zero (that is, to fractions), e.g. in the number 347.12, 1 represents one-tenth, 2 represents two-hundredths. The dot between the whole number part and the fractional part is the decimal point. In some countries (Canada, USA) this point is placed on the line, in some (UK) it is placed above the line (347·12), and in some (France) a comma is used instead (347,12).

If there is no whole number part, it is usual to mark this with a zero, e.g. 0.12 is a decimal fraction, meaning 12 hundredths.

The number 47.12 is read as either 'forty-seven point one two' or as 'forty-seven and twelve hundredths'.

A decimal number is sometimes called simply a decimal.

Many countries have adopted a decimal currency (one based on 10s) for their money system.

Example
        USA, 1 dollar = 100 cents
        France, 1 franc = 100 centimes
        UK, 1 pound = 100 pence
        China, 1 yuan = 100 fen
        Canada, 1 dollar = 100 cents

The metric system of measures is a decimal system.

LATIN   *decimus*: tenth

decreasing (a)   Growing less.

y is said to be a decreasing **function** of x if y continuously grows less as x grows larger (see Figure D1).

LATIN   *decrescere*: to grow less

Figure D1 graph of a decreasing function

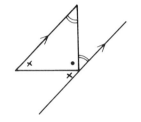

Figure D2

deduction **1** Taking away. Subtraction.
2 Drawing conclusions by reasoning from facts or principles.
Example
One method of proving that the three angles of a triangle add up to two right angles is by deduction from the angle properties of parallel lines, as suggested in Figure D2.
3 The conclusion reached by the method of deduction.
LATIN *deducere*: to lead away

definition [def-in-*ish*-un] An exact description, using terms already described.
  The formal definition of an object in mathematics states the least number of properties of the object needed to identify it.
Example
If a **quadrilateral** has already been defined as a plane four-sided figure, then the definition of a **parallelogram** may be expressed as 'a quadrilateral with both pairs of opposite sides parallel'. By using the term 'quadrilateral' in the definition it does not have to be stated that a parallelogram is also a four-sided plane figure. Further, it can be shown that the opposite sides and angles are equal, but these properties are not part of the definition.
LATIN *definere*: to set bounds

degree
  1 (algebra) — The **power** of a term or expression.

Example
x^3, x^2y and $5x^2y$ are terms of the 3rd degree. $2x^3 + 3x^2 - x + 1$ is an expression of the 3rd degree.
2 (geometry) — The size of an angle that is 1/360th of a complete rotation (sometimes referred to as a degree of arc). One-quarter of a rotation (a right angle) is ninety degrees, written 90°.
3 (measurement) — A unit for measuring temperature. The boiling temperature of water is 100 degrees on the Celsius scale (written 100°C) and 212 degrees on the Fahrenheit scale (written 212°F).
LATIN   *de*: from, *gradus*: a step

deka-   Used (but rarely) in the international system of units (SI) to denote ten times. Symbol da.
Example
$$1 \text{ dekalitre} = 10 \text{ litres}$$
$$1 \text{ daL} = 10 \text{ L}$$
GREEK   *deka*: ten

delta   Fourth letter of the Greek alphabet, Δ and δ. Used in mathematics to mean 'a small change in the value of a variable'.
Example
δx or Δx means 'a small increase in the value of x'. If x changes from $x = 3$ to $x = 3.1$, then $\delta x = 0.1$.

denary (a) [*dee*-na-ree]   Based on ten. The denary number system is the same as the decimal number system.
LATIN   *denarius*: containing ten

denominator [dee-*nom*-in-ay-tor]   Of a **fraction** written in the form $\frac{a}{b}$, the number below the line. It is the **divisor** of the number above the line (the **numerator**).
Examples
The denominator of the fraction $\frac{3}{4}$ is 4; the

denominator of $(x^2 - 1)/(x + 1)$ is $x + 1$.
See also **least common denominator**.
LATIN *denominare*: to name

density [*den*-si-tee]   Mass per unit volume of a substance.
Example
Every cubic metre of pure water has a mass of 1000 kilograms. Therefore the density of water is said to be 1000 kilograms per cubic metre, written 1000 kg.m^{-3}.

dependent (a) [de-*pen*-dent]   Relying on another.
   In mathematics, the value of a dependent variable depends on the values given to another variable (the **independent** variable).
Example
In the equation $y = x^2$, the values of y depend on the values chosen for x; x is the independent variable and y is the dependent variable. To show the relationship on a graph, the independent variable is usually plotted horizontally. The dependent variable is plotted vertically (see Figure D3).

x	−2	−1	0	+1	+2
y	+4	+1	0	+1	+4

LATIN *dependere*: to hang down

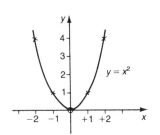

Figure D3

derivative [de-*riv*-a-tiv]   An important type of function used in **calculus** for calculating the rate of change of a given function. It is also called derived function and differential coefficient.

Descartes [day-*kart*] **René** (1596–1650)   A famous French philosopher and mathematician who lived much of his life in Holland and other European countries. His most important contribution to mathematics is his method of relating geometry and algebra. For instance, the geometric curve known as a **parabola** can be expressed algebraically as an equation of two variables, e.g. $y = ax^2$, and pairs of values of the algebraic variables, x,y, in the equation $y = ax^2$ can be interpreted as points on a parabola.

In general, Descartes established a **one-to-one correspondence** between plane curves in geometry and algebraic equations of two variables—for each curve there is an equation in two variables; for each equation in two variables there is a plane curve. As part of this work, he invented the system of **Cartesian** coordinates still in use today.

Descartes also saw that his method can be extended to three dimensions, relating **surfaces** to equations in three variables.

describe (v) To mark out or draw, e.g. a pair of compasses is used to describe a circle.
LATIN *de-*: down, *scribere*: to write

determine (v) To specify precisely.
Example
Two points determine a line, in the sense that there is one line and only one line that can be drawn through two given points. Similarly, three points that are not in line determine a circle, since there is a single circle that can be drawn to pass through them.
LATIN *de-*: from, *terminare*: to limit

develop (v) In geometry, to roll out a surface on to a plane.
Example
If the curved surface of a right circular **cylinder** is rolled out on a plane, the result is a **rectangle**. Figure D4 shows one way of developing a **cube**. Not all surfaces can be developed, e.g. it is not possible to develop a sphere.
See also **net**.
FRENCH *développer*: to unfold

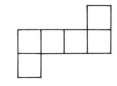

Figure D4

deviation In a set of numbers, the difference between one number and the **mean** of the set.

Example
For the following scores in a diving competition:
4.3, 6.8, 6.3, 4.7, 4.5, 6.4, 5.5, the mean is 5.5.
The deviations from this mean are: −1.2, +1.3, +0.8, −0.8, −1.0, +0.9, and 0. Note that the deviations always add up to zero.
The mean deviation is calculated by disregarding the signs: (1.2 + 1.3 + 0.8 + 0.8 + 1.0 + 0.9 + 0) ÷ 7 ≈ 0.9. This is sometimes used as a measure of **dispersion**.
See also **standard deviation**.
LATIN *deviare*: to turn away

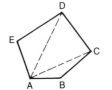
Figure D5

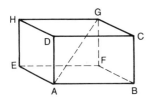
Figure D6

diagonal [dy-*ag*-on-l] A line segment joining two non-adjacent **vertices** of a **polygon** or two vertices of a **polyhedron** not on the same **face**.
Examples
a In Figure D5, AC and AD are two diagonals of pentagon ABCDE.
b In Figure D6 AG is a diagonal of a cuboid. AH and AC are diagonals of faces of the cuboid but not of the cuboid as a whole.
c A polygon with *n* sides has altogether $\frac{1}{2}n(n-3)$ diagonals.
GREEK *dia-*: through, *gonia*: angle

diagram A line drawing, graph or picture used to help with explanations in mathematics.
GREEK *dia-*: through, *gramma*: written

diameter [dy-*am*-e-ter] A line segment joining two points on the **circumference** of a circle and passing through the centre. All the diameters of a circle are equal in length and twice as long as a **radius**.
The term is also used in a similar way in connection with a sphere.
GREEK *dia-*: through, *metron*: measure

diametrical (a) [dy-a-*met*-rik-al)] Relating to a diameter, or along a diameter.

diamond A **rhombus**, i.e. a **parallelogram** with adjacent sides equal.

Dido's problem [*dy*-doh] According to an ancient Greek legend, Dido became the first queen of Carthage in North Africa when a storm drove her fleet ashore there. The inhabitants allowed her to buy as much land as could be enclosed by a bull's hide. This turned out to be quite a large area, because she cut and twisted the hide into a long cord and then formed a circle with it. This is an illustration of the fact that, for a given length of **perimeter**, the shape that encloses the largest area is a circle.

die (*pl.* **dice**) A small **cubic** solid made of wood, bone, plastic, etc., each face being marked with a different number of spots. The most usual arrangement is for each face of a die to have 1,2,3,4,5 or 6 spots with opposite faces having 1 and 6, 2 and 5, 3 and 4 respectively. Dice are used in games of chance, and in mathematics to illustrate the laws of **probability**.

difference The amount by which a quantity is more or less than another.
Examples
The difference between 80 and 120 is 40; the difference between -10 and -12 is 2; the difference between -2 and 3 is 5.

difference of two squares In algebra, an expression of the form $a^2 - b^2$.
Example
Each of the following can be expressed as the difference of two squares:
a $4x^2 - 9y^2 = (2x)^2 - (3y)^2$
b $100k^2 - 1 = (10k)^2 - (1)^2$
To recognise a difference of two squares is helpful in factorising some algebraic expressions, because $a^2 - b^2$ has the **factors** $(a - b)(a + b)$.

differential calculus That part of **calculus** that deals with rates of change. The other main part is **integral** calculus.

digit [*dij*-it] Any one of the basic counting symbols in a number system.
Example
In the **decimal** system, the digits are 0,1,2,3,4,5,6,7,8,9. In the **binary** system, the digits are 0,1.
See also **Arabic numerals**.
LATIN *digitus*: finger or toe

digital (a) [*dij*-it-l] Referring to machines and devices that hold or process information in the form of digits. Most often the digits are 0, 1 representing 'off' and 'on' for an electronic circuit. The development of digital devices such as computers and compact discs has emphasised the importance of understanding the mathematics of number systems.

In contrast, **analog** devices use physical quantities instead of digits for storing and processing information.

dihedral (a) [dy-*hee*-drl] Referring to two planes.

A dihedral angle is the angle between two planes. To find this angle, e.g. between the pages of a partly open book, mark a point on the line along which the two planes meet and draw two other lines perpendicular to this line of intersection, one in each plane. The angle between the two perpendiculars is the dihedral angle (see Figure D7).
GREEK: *dis*: twice, *hedra*: base

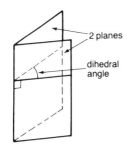
Figure D7

dilatation A transformation of a geometrical figure carried out in such a way that the new form has a **similar** shape to the original. It may be larger or smaller.

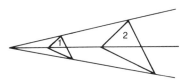

Figure D8

Example
Triangle 1 in Figure D8 has been enlarged to triangle 2 by a process of positive dilation, or triangle 2 has been reduced to triangle 1 by a process of negative dilation. Notice how the lines joining corresponding points in the two figures are **concurrent**.

The ratio of lengths measured on the new form to corresponding lengths on the original is called the **scale factor**; a scale factor of 2 means that lengths have been doubled.

Also called dilation.

LATIN *dilatare*: to spread out

dimension

1 The dimensions of a point in geometry are the number of **coordinates** needed to specify its position. A point that can move only along a certain line needs only one coordinate to specify its position (which could be its distance from a fixed point on the line). A point on a plane needs two coordinates (which could be its distances from two intersecting lines in the plane, as in the **Cartesian** system). A point in ordinary space needs three. So it is said that a line is one-dimensional, a plane is two-dimensional, and ordinary space is three-dimensional.

The regions enclosed by triangles, polygons, circles, etc. are two-dimensional because they are parts of planes. Regions enclosed by cubes, cones, cylinders, etc. are three-dimensional.

In advanced mathematics, the properties of n-dimensional space (where $n > 3$) are examined.

2 The dimensions of an object express its length, area or volume, e.g. the dimensions of a room may be 5 metres wide × 6 metres long × 2.6 metres high.

3 The term is also applied to physical quantities themselves, such as length, volume, velocity, force, energy. The fundamental

direct proportion

dimensions are **length** (L), **mass** (M) and **time** (T). Other dimensions can be expressed in terms of these three, e.g. velocity, being measured as length travelled in unit time, has the dimensions LT^{-1}.

LATIN *dimetiri*: to measure out

direct proportion Two variables are in direct proportion if the **ratio** of one to the other remains constant. Also called direct **variation**.
Example

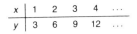

In the table of values given in Figure D9, y/x remains constant at value 3, so x and y are in direct proportion.
The general equation for two variables x, y that are in direct proportion is $y = kx$ where k is a constant. The graph of y against x is a straight line passing through the origin.

The following are equivalent statements:

x and y are in direct proportion.
y is directly proportional to x.
y varies as x.
$y \propto x$.
$y = kx$ where k is a constant.

See also **inverse proportion**.

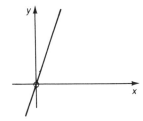

Figure D9

directed number A number that has a positive or negative sign. Directed numbers are useful in the measurement of quantities that have the idea of greater or less than some reference value.
Examples
Temperature may be above or below the freezing point of water; geographical height may be above or below sea level; a bank balance may be in credit or in debit. The sense of greater than/less than, above/below, etc. is conveyed by the use of + or − prefixed to the numbers, and the numbers can be represented on a number line with negative values to the left of zero and positive values to the right:

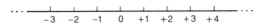

+3 is read as 'positive 3' or, less strictly, 'plus 3'.
−2 is read as 'negative 2' or, less strictly, 'minus 2'.

(Strictly speaking, 'plus' and 'minus' refer to the operations of addition and subtraction, not to positions on a number line, but often this distinction is not made. Also, some teachers write $^+3$ and $^-2$ for positive 3 and negative 2, reserving + and − in the normal position for addition and subtraction.)

discontinuity A value for x on an x–y graph for which a value for y is not defined.
Example
The graph of $y = \tan x$ has a discontinuity at $x = \pi/2$ (see Figure D10).
See also **continuous**.
LATIN *dis*: away, *continere*: continue

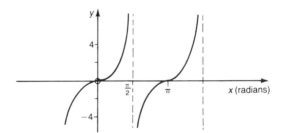

Figure D10

discount An amount to be subtracted from the listed price of something.
Example
For a $20 article sold at $2 discount, a buyer would pay $18.

Discounts are often expressed as a percentage of the selling price. The above example illustrates a discount of 10%, and the buyer would pay 90% of the listed price.
LATIN *dis*: away, *computare*: to calculate

discrete (a) Consisting of separate and distinct parts. **Discrete** variables measure things that can be counted using whole numbers, such as books on a shelf, people in a room, letters in a word. **Continuous** variables are not discrete, e.g. the temperature of a room.

LATIN *discretus*: separated

discriminant [dis-*krim*-in-ant] An expression occurring in the theory of **quadratic equations**.

The discriminant of the equation $ax^2 + bx + c = 0$ is $b^2 - 4ac$. Its value determines whether the equation has:

two real roots $(b^2 - 4ac > 0)$
one real root $(b^2 - 4ac = 0)$
no real roots $(b^2 - 4ac < 0)$.

Example
The discriminant of the equation
$x^2 - 6x + 5 = 0$ is
$(-6)^2 - 4 \times 1 \times 5 = 36 - 20 = 16$. The discriminant is thus greater than zero and the equation has two real roots. (They are $x = 5$ and $x = 1$.)

disjoint (a) Of sets, having no elements in common.

Example
The set of even integers and the set of odd integers are disjoint, because an even integer cannot be odd, and an odd integer cannot be even.

The following two sets are not disjoint, since 6 is common to both: A = {2,3,6,8}, B = {3,6,9,12}. A ∩ B = {6}.

The **intersection** of two disjoint sets is always the **empty** set, ∅.

LATIN *disjungere*: to separate

disjunction
1 In logic, a **connective**, the equivalent of 'or' that combines two **propositions**. Its symbol is v.
2 In logic, the combining of two propositions

with 'or'. In this sense, pvq is a disjunction formed from the two propositions p and q.
Example
If p is the proposition 'I am in London' and if q is the proposition 'Today is Thursday', then pvq represents the proposition 'I am in London or today is Thursday'. (It is important to note that this use of the word 'or' means that either of the original propositions may be true or that both may be true. This is called the inclusive use of 'or'.)
See also **conjunction, truth table.**
LATIN *dis-*: apart, *jungere*: to join

dispersion The extent to which the values in a **frequency distribution** are spread about a central value (usually the **arithmetic mean**).
Example
The following two sets of numbers have the same mean (15), but the numbers in the first are more closely packed around the mean and so the dispersion is less:
    8, 10, 10, 12, 15, 17, 19, 20, 24
    8, 8, 8, 9, 15, 20, 21, 22, 24
Sometimes the difference between first and last value (the **range**) is taken as a measure of dispersion; but in the example above this would not distinguish between the two distributions. Better measures of dispersion are **interquartile range** and **standard deviation**.
See also **variance.**
LATIN *di-*: apart, *spargere*: scatter

displacement Change of position, described as both distance and direction from the starting point.
Example
Someone standing on top of a wall 20 metres high throws a ball upwards. If the ball then falls down to earth, its final displacement from the top of the wall is described as 20 metres downwards, even though it has actually travelled

distance

more than 20 metres altogether.
Displacement is a **vector** quantity, whereas distance is a **scalar** quantity.
LATIN *dis*: apart, **platea**: a place

distance The separation between two things measured in units of length, or the length of a path joining two points.
Examples

Figure D11

a The distance between two points is measured along the straight line joining the two points. In Figure D11, the distance between P and Q = length *PQ*.

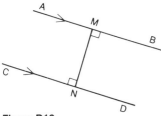

Figure D12

b The distance between two parallel lines is measured along a line perpendicular to both. In Figure D12 the distance between AB and CD = length *MN*.

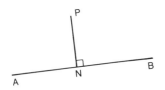

Figure D13

c The distance between a point and a line is measured along the perpendicular drawn from the point to the line. In Figure D13, the distance between P and AB = length *PN*.

d The distance between two points on a curved surface (as on the earth) may be measured along the shortest curve joining the two points. This curve is called a **geodesic** (*see also* **great circle**).

There are different definitions for distance in **non-Euclidean geometry**.

distribution A set of observations or measurements.
Example
An observer on a street corner took note of the number of people in the cars that passed in 5 minutes:
   1, 1, 3, 1, 4, 5, 1, 2, 2, 1, 3, 4, 1, 2, 1.
This set of numbers is a distribution. Regrouped as follows, it becomes a **frequency distribution**:
   number of persons 1   2   3   4   5
   number of cars   7   3   2   2   1
LATIN *distribuere*: divide

divisibility

distributive law [dis-*trib*-yu-tiv] or **property**
The rule that states the truth of the formula:
$$a \times (b + c) = a \times b + a \times c$$
Multiplication is said to be distributed over addition.
Division is not distributed over addition:
$a \div (b + c)$ is not the same as $a \div b + a \div c$
LATIN *distribuere*: divide

divergent (a) [dy-*ver*-jent]
1 (geometry) — Aiming in different directions from a point, e.g. the rays OA and OB in Figure D14 are divergent.
2 (algebra) — Describing a **series** whose sum has no limit as the number of terms increases.
Examples
a $1 + 2 + 4 + 8 + \ldots$ is divergent.
b $1 + \frac{1}{2} + \frac{1}{3} + \frac{1}{4} + \ldots$ is also divergent, but
c $1 + \frac{1}{2} + \frac{1}{4} + \frac{1}{8} + \ldots$ has a sum that never grows greater than 2; it is **convergent**.
LATIN *divergere*: to turn away

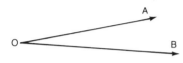

Figure D14

dividend [*div*-i-dend] 1 A number being divided. In the example, $18 \div 3 = 6$, 18 is the dividend, 3 is the divisor, 6 is the quotient.
2 A share of profits distributed by a company to its share-holders.
LATIN *dividendum*: to be divided

divisibility [di-viz-i-*bil*-i-tee]   The property of a number that it is able to be divided exactly by another.
  Here are some useful tests for the divisibility of whole numbers:
- All numbers ending in 0,2,4, 6,8 are divisible by 2.
- If the last two digits form a number divisible by 4, then the total number is divisible by 4, e.g. 96 is divisible by 4, so also are 1996, 7396, 432 096, etc.
- If the last three digits form a number

divisible by 8, then the total number is divisible by 8.
- All numbers ending in 0 or 5 are divisible by 5.
- All numbers ending in 0 are divisible by 10.
- If the digits add up to a number divisible by 3, then the original number is divisible by 3, e.g. 3825: $3 + 8 + 2 + 5 = 18$, which is divisible by 3, so 3825 is also divisible by 3.
- If the digits add up to a number divisible by 9, then the original number is divisible by 9.

divisible (a) [di-*viz*-i-bl]   Describing a number that exactly contains another number repeatedly, without remainder.
Example
18 is divisible by 3 (6 times exactly), but 19 is not divisible by 3 (6 times and a remainder of 1).
See also   **divisibility**.

division   The **inverse** of **multiplication**.
a divided by b is written as $\frac{a}{b}$ or a/b or $a \div b$.

Rules for the division of **signed numbers**:
- If the two numbers have the same sign, the result of division is positive, e.g.
$+8 \div +2 = +4; -10 \div -2 = +5$
- If the two numbers have different signs, the result of division is negative, e.g.
$+8 \div -2 = -4; -10 \div +2 = -5$

Rules for zero:
- 0 divided by any other number gives the result 0, but division by 0 is not allowed, e.g.
$0 \div 12 = 0$; but $12 \div 0 = ?$

Division of one whole number by another whole number does not always give a whole number, e.g. $10 \div 4 = 2.5$. This shows that the set of whole numbers is not **closed** for division (whereas it is closed for multiplication).

divisor [di-*vy*-zer]   A number by which another is divided. In the example, $18 \div 3 = 6$, 3 is the divisor, 18 is the dividend, 6 is the quotient.

dodeca- Prefix meaning twelve (*see* next entry).
GREEK   dodeka: twelve

dodecagon [doh-*dek*-a gon]   A twelve-sided **polygon**. Each **interior** angle of a **regular** dodecagon measures 150°.
GREEK   *dodeka*: twelve, *gonia* angle

dodecahedron [doh-dek-a-*hee*-dron]   A twelve-faced **polyhedron**. It is possible to construct a **regular** dodecahedron; its faces are regular pentagons.
GREEK   *dodeka*: twelve, *hedra*: base

domain   The set of possible values for which a **function** exists.
Example
The function $\sqrt{x}$ in ordinary algebra exists only for non-negative values of x. The domain of this function consists therefore of zero and all positive numbers.

On a **Cartesian** graph, the domain is the set of values of x (the **independent** variable) for which there are values of y (the **dependent** variable) (see Figure D15). The corresponding values of y make up the **range** of the function.
LATIN   *dominus*: master

Figure D15

dot plot — see **line plot**.

duodecimal (a)   Belonging to a system of counting by twelves.

Duodecimal reckoning is still familiar today in the use of the terms 'dozen' (12) and 'gross' (12 × 12); in the division of day and night into 12 hours each; and in the old units of length (12 inches to a foot) and of money (12 pence to a shilling). Note also that the words 'eleven' and 'twelve' are not like 'thir*teen*', 'four*teen*', etc., which contain a suggestion of *ten*.
LATIN   *duodecim*: twelve

dynamics [dy-*nam*-iks]   The study of forces acting on a body to produce motion or change of motion.
Compare **kinematics, statics**.
GREEK   *dynamis*: power

E

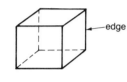

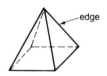

Figure E1

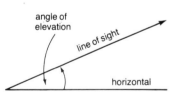

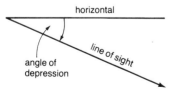

Figure E2

e   The **base** of the Naperian or natural system of **logarithms**. The Naperian logarithm of a number x is written $\log_e x$ or $\ln x$. e is an **irrational** number with the approximate value 2.718. Its value can be calculated to any desired accuracy by adding together terms of the form $1/n!$ for values $0,1,2,3,\ldots$ of n:
$$e = 1 + 1 + \tfrac{1}{2} + \tfrac{1}{2\times 3} + \tfrac{1}{2\times 3\times 4} + \ldots$$
The use of the letter e for this purpose was suggested by **Euler** in the 18th century.

edge   The line segment where two plane faces of a solid figure meet, e.g. a cube has twelve edges, a square pyramid has eight.

element   One of the objects or numbers belonging in a **set**.
The symbol ϵ means 'is an element of' e.g. if set P is $\{3,6,9,12\}$, then $6 \in P$.

elevation [el–ev-*ay*-shun]   **1** A drawing to scale of a solid object as seen from the front or side. *See also* **plan**. **2** The angle between the horizontal and the line of sight to an object. If the object is below the horizontal, the angle is referred to as an angle of depression (see Figure E2).
LATIN   *elevare*: to lift up

eliminate (v) [e-*lim*-in-ayt]   To remove a **variable** from **equations** in algebra.
Example
y may be eliminated from the following pair of **simultaneous** equations:
$$2x + 3y = 13$$
$$3x - y = 3$$

To do this, multiply all the terms of the second equation by 3 and then add to the first:

2x + 3y = 13
9x − 3y = 9
11x   = 22   y is now eliminated.

LATIN *eliminare*: to banish

ellipse [ee-*lips*] The path traced out by a point moving in a plane so that the sum of its distances from two fixed points remains constant.

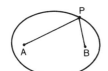

Figure E3

In Figure E3, A and B are the fixed points (each is called a **focus** of the ellipse), and P is the moving point, which traces out an ellipse according to the rule $AP + BP =$ constant. The closer that A and B are together, the more nearly does the ellipse look like a circle. Its difference from a true circle is known as its eccentricity. A circle may be thought of as a special ellipse with zero eccentricity.

An ellipse is also the shape of some cross-sections of a **cone** (see Figure E4). In forming an ellipse this way, a plane cuts the cone at an angle that is less than the angle that the side makes with the base. Another way of looking at this is that an ellipse results from the **projection** of a circle onto a non-parallel plane.

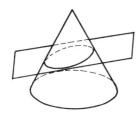

Figure E4

An ellipse is symmetrical about two axes: a major axis and a minor axis (see Figure E5). *See also* **conic section**.

GREEK *elleipsis*: defect or omission

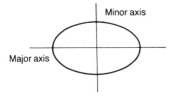

Figure E5

ellipsoid A solid figure formed by rotating an **ellipse** about its major axis or its minor axis. A cross-section perpendicular to the axis of rotation is a circle; one parallel to the axis is an ellipse.
Also called **spheroid**.

empirical (a) [em-*peer*-i-kl] Referring to evidence obtained by observation or experiment, not deduced by logic or theory.

Example
Some probabilities can be calculated theoretically. For example, the **probability** of rolling a 5 with a die is 1/6 because only one face of the die has five spots and there are six faces. Some other probabilities can be calculated only by empirical means, that is, by observing past events. For example, to estimate the probability that the next baby born at a certain hospital will be a girl, the records for the past 10 years, showing that 481 out of every 1000 births were girls, could be used. Therefore the empirical probability of a girl being born next is 48.1%.

GREEK *empeirikos*: a physician guided by experience

empty (a)   Having no members or elements.
In **set** theory, the empty set is represented by { } or ∅. It is also called the null set. Note that there is only one empty set. Neither a set of all square circles nor a set of unicorns has any members; they are therefore identical and described as the empty set.

enantiomorphic (a) [en-an-ti-oh-*mor*-fik]   Describing two **asymmetric** figures that are the mirror image of each other. Like a pair of gloves.

GREEK *enantio*: opposite, *morphe*: form

enlargement   *See* **dilatation**.

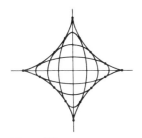

Figure E6

envelope [*en*-vel-ohp]   A curve that is **tangent** to every member of a **family** of curves.
Example
If a family of ellipses is drawn such that the sum of their minor and major axes remains constant, then the envelope is an **astroid** (see Figure E6).
In three dimensions, an envelope is a surface that is tangent to a family of planes or surfaces.

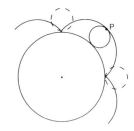

Figure E7

epicycloid [ep-ee-*sy*-kloyd] The path traced out by a point on a circle as the circle rolls around the circumference of another circle in the same plane. Figure E7 shows part of an epicycloid traced out by the point P on the smaller circle.

The **cardioid** is an example of an epicycloid.
See also **cycloid**.
GREEK *epi-*: on, *kyklos*: circle

epsilon [*eps*-i-lon] Fifth letter of the Greek alphabet, written E, ϵ, corresponding to English short E, e. ϵ is used sometimes in mathematics to represent a small positive quantity, and also to mean 'is an **element** of the set'.

equal (a) Alike in some way.
Example
7 + 5 and 4 × 3 are equal in value.
 The two sets {a,b,c,} and {12,13,20} are equal in number of elements.
 Signs used:
  = 'is equal to' or 'equals'   3 × 4 = 12
  ≠ 'is not equal to'   3 × 4 ≠ 11
  ≐ 'is approximately equal to' $\sqrt{2} \doteq 1.414$
LATIN *aequalis*: equal, like

equation [ee-*kway*-shun] A statement that two mathematical expressions have the same value, i.e. they stand for the same number.
Example
$5a = 30$ is an equation. It states that $5a$ and 30 have the same value. This is true only if a is equal to 6, so $a = 6$ is called the **solution** (or **root**) of the equation.
 An equation that is always true is called an **identity**, e.g. $2(a + 4) = 2a + 8$ is true for all values of a.
See also **equation of a line or curve, quadratic equation, simultaneous equations**.
LATIN *aequatus*: made equal

equilateral

equation of a line or curve An algebraic statement that defines all the points on a line or curve. If the line or curve is drawn with **Cartesian** axes, the equation is a statement about the x- and y- **coordinates** of the points. Three examples are shown in Figure E8.

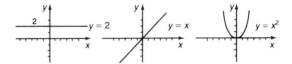

Figure E8

equator [ee-*kway*-tor] A circle that divides a sphere or other surface into two equal and symmetrical parts. The earth's equator is the **great circle** half-way between the north and south poles. A place on the earth's equator has **latitude** 0°.
LATIN *aequator*: equaliser

equiangular (a) [ee-kwi-*ang*-yoo-lar] Having all angles equal.
Example
A regular **hexagon** has all six angles equal to 120°. It is therefore equiangular.
   An **equilateral** triangle is equiangular; other triangles are not.

equidistant (a) Being the same distance away.
Example
The centre of a circle is equidistant from all points on the circumference; but there is no point equidistant from all points on an ellipse (see Figure E9).

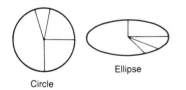

Circle
Ellipse

Figure E9

equilateral (a) [ee-kwi-*lat*-er-al] Having all sides equal in length.
Example
An equilateral triangle has three equal sides. It is also **equiangular**. Other equilateral plane

figures may or may not be equiangular. A **square** and a **rhombus** are equilateral, but only the square is equiangular.
LATIN *aequus*: equal, *latus*: a side

equivalent (a) [ee-*kwiv*-a-lent] Equal in value.
Examples
a Equivalent **fractions** are fractions that can be reduced to the same number, e.g. all these fractions can be reduced to $\frac{1}{2}$ and so are equivalent: 2/4, 3/6, 50/100, 0.5, 50%
b Equivalent **expressions** are expressions that are always equal to each other, e.g. $2(a - 5)$ and $2a - 10$ are equivalent expressions.
c Toplogically equivalent — see **topology**.
See also **identity**.
LATIN *aequus*: equal, *valens*: power

Eratosthenes [e-ra-*tos*-the-neez] A Greek scholar who lived in the 3rd century BC. In the field of mathematics, he measured the length of a degree on the earth's surface, reformed the calendar, and showed how to sift the **prime** numbers from the rest of the whole numbers.

error The difference between the true value of a quantity and an approximate measurement of it. It is not always possible to know the true value, so the error is estimated.
Example
If the width of a book is measured with a ruler marked in millimetres, the result may be shown as 18.4 ± 0.05 centimetres. This means that the true value lies close to 18.4 centimetres with a likely error of no more than 0.05 centimetres ($= \frac{1}{2}$ millimetre) either way.
See also **accuracy**, **approximation**.
LATIN *errare*: to wander

escribed circle A circle drawn so that one side of a triangle and extensions of the other two

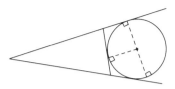

Figure E10

sides are **tangents** to it (see Figure E10). Every triangle can have three escribed circles. Also known as ecircle and excircle.

LATIN   *ex-*: out of, *scribere*: to draw

estimate (v)[*es*-ti-mayt]   To judge or calculate approximately the amount or value or size of something.

estimate (n) [*es*-ti-met]   a number or quantity serving as an approximation to the true value. Estimates are useful for checking exact calculations.
Example
   calculation $41.25 \times 2.9 = 1196.25$
   estimate $40 \times 3 = 120$
Comparison shows up an error in the first line where the result should be 119.625

Euclid [*yoo*-klid]   A Greek mathematician who lived in Alexandria during the 3rd century BC. His most famous written work, *Elements*, presented a systematic treatment of all known geometry. In it, he attempted to show that all geometrical propositions could be proved on the basis of a few assumptions called **axioms**. Much of Euclid's geometry has continued to be taught in schools right up to the 20th century.

Euclidean (a) [yoo-*klid*-i-an]   An adjective applied to ordinary geometry based on **Euclid**'s assumptions. Until the 19th century it was the only geometry, but there are now other kinds of geometry based on different assumptions about **parallel** lines and about **distance**.
See also   **non-euclidean geometry, space**.

Euler [*oy*-ler], **Léonard** (1707–1783)   Of Swiss nationality, one of history's most important mathematicians. His work covered many branches of mathematics, including **calculus, circular functions, complex numbers, primes**, as well as the application of mathematics to astronomy, science and engineering. He standardised the use of several symbols: f(x) for **function**, Σ for **sum**, **i** for $\sqrt{-1}$, **e** for the base of natural logarithms, π for the ratio of circumference to diameter of a circle.

Euler diagram A diagram (often consisting of circles) used in logic to represent the relation between **sets**. Figure E11 represents the possible types of relation between two sets.
See also **venn diagram**.

P and Q **intersect**   P includes Q   P and Q are **disjoint**

Figure E11

Euler's formula A formula relating the numbers of **vertices** (V), **faces** (F) and **edges** (E) of a **polyhedron**: $V + F - E = 2$.
Examples
For a square pyramid, $V = 5$, $F = 5$, $E = 8$; for a cube $V = 8$, $F = 6$, $E = 12$.

The formula is also adapted to **networks** in **topology** in the form $N + R - A = 2$, relating the number of nodes (N), regions (R) and arcs (A), e.g. in Figure E12 there are 5 nodes, 4 regions, 7 arcs, and $5 + 4 - 7 = 2$.

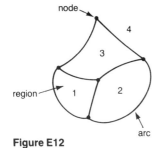

Figure E12

evaluate (v) [ee-*val*-yoo-ayt] To find a particular **value** of a **function** or **expression**.
Example
To evaluate $5x^2 + 1$ for $x = 3$, means to substitute 3 for x and get the result $5 \times 3^2 + 1 = 46$.
LATIN   *e*-: from, *valere*: to be worth

even (a)
1   Even number. A whole number exactly divisible by 2. Even numbers are **multiples** of 2

and belong to the sequence ... −4, −2, 0, +2, +4, +6, ... The other whole numbers are **odd** numbers. Even numbers added together or multiplied together produce even numbers, e.g. $8 + 4 = 12$; $6 \times 4 = 24$. An even number added to an odd number produces another odd number, e.g. $4 + 3 = 7$. An even number multiplied by an odd number produces an even number, e.g. $4 \times 3 = 12$.

2 Even function. A **function** whose graph is **symmetrical** about the y-axis (see Figure E13).

If f(x) is an even function, then replacing x by $-x$ does not change its value: $f(-x) = f(x)$.
Example
x^2 is an even function because $(-x)^2$ has the same value as x^2. x^3 is not an even function because $(-x^3)$ has the opposite sign from x^3.

3 Even chance. A **probability** that is the same for an event to happen as for it not to happen, e.g. if a coin is tossed, there is an even chance that it will come down heads.

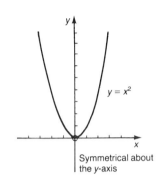

Figure E13 — Symmetrical about the y-axis

event In statistics, an experiment or a possible observation or outcome of an experiment.
Example
The throw of a die is an event and each of the possible outcomes (1,2,3,4,5 or 6) is an event. If two coins are tossed, there are four possible events: HH, HT, TH, TT.

Events may be **mutually exclusive** or **independent**.
See also **compound event**.

evolution [ev-ol-*yoo*-shun] In algebra, the operation of finding the **root** of a number. Finding the square root and the cube root are examples of evolution. The inverse operation is called **involution**.
LATIN *evolvere*: to roll out

exa- A prefix meaning a million million million times. Symbol E

Example
1 exatonne = 1 000 000 000 000 000 000 tonnes
1 Et = 10^{18} t

expand (v)   To write an expression in extended but equivalent form.
Example
$(a + b)^2$ may be expanded to $a^2 + 2ab + b^2$.
LATIN   *e-*: out, *pandere*: to spread

expansion   A longer form, equivalent to a given expression.
Example
$x^3 + 3x^2 + 3x + 1$ is an expansion of $(x + 1)^3$
LATIN   *e-*: out, *pandere*: to spread

expected value   The theoretically calculated outcome for a set of observations based on **probability**.
Example
Because the probability of one toss of a coin resulting in a head is $\frac{1}{2}$, the expected value for 100 tosses is 50 heads. In an actual experiment, the number of heads may be more or less than 50.
See also   **law of large numbers**.

explicit (a)   [eks-*plis*-it]   Stated directly.
Example
$y = 2x + 1$ states y directly as a function of x; y is an explicit function of x. Contrast this with $x^2 + y^2 = 4$. In this case, y can be regarded as an **implicit** function of x.
LATIN   *e-*: out, *plicare*: to fold

exploded drawing   A drawing showing all the parts of an object laid out separately, though in proper relation to each other (see Figure E14).

exponent   [eks-*poh*-nent]   A number expressing a **power**.

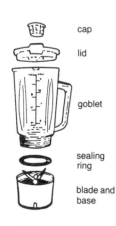

cap
lid
goblet
sealing ring
blade and base

Figure E14

Example
In $3^2 = 9$, 2 is the exponent and 3 is the base. An exponent is often referred to as an **index**.
LATIN   *ex-*: out, *ponere*: to place

exponential function [eks-po-*nen*-shul]   A function (y) of the form $y = a^x$, where a is a positive constant.

For values of a greater than 1, the value of the function rises more and more rapidly as x increases. For values of a less than 1, the value of the function falls less and less rapidly as x increases (see Figure E15).

Exponential functions are useful in describing growth, e.g. unrestricted population growth, and decay, e.g. decay of a sample of radioactive material.

The most fundamental exponential function in mathematics is $y = e^x$, where e is the constant 2.718... (*see* **e**). It has the special property that, at any point on its graph, the **gradient** has the same value as x.

The **inverse** of an exponential function is a **logarithmic function**, e.g. the inverse of $y = 10^x$ is $x = \log_{10} y$.

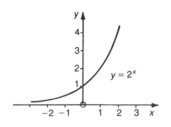
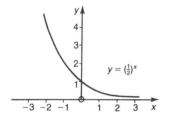

Figure E15

exponential series   The series, $1 + x + x^2/2 + x^3/6 + \ldots$ Each term in the series is of the form $x^n/n!$.

This is an important series in **complex number** theory. It also provides a way of calculating the value of **e**. (Put $x = 1$ in the series.)

expression   A general term in mathematics for a collection of symbols representing numbers and operations.
Examples
$5 - \sqrt{3}$; $x^2 - 2x + 1$; $1 + \log x$.

An algebraic expression is one that uses only **algebraic operations**.
LATIN   *exprimere*: to squeeze out

exterior

Figure E16

exterior (a) Outside. Opposite of **interior**. An exterior angle of a triangle or polygon is formed when a side is **produced** as in Figure E16. An exterior angle and the **adjacent** interior angle add up to 180°.

LATIN *exter*: on the outside

extract (v) To find the value of. Used mostly in connection with **roots**, e.g. to extract the square root of 2 means 'to calculate $\sqrt{2}$'.

LATIN *ex-*: out of, *trahere*: to pull

extrapolate (v) [eks-*trap*-o-layt] To estimate a value by following a pattern and going beyond values already known.

Example

In an experiment on stretching, weights were added to a spring and the following measurements were obtained:

Weight added (g)   20 40 60 80 100 120 140
Length of spring (cm) 14 21 28 35 41   46   50

It was noticed that, after the first few weights, the length of the string increased by 1 centimetre less for each 20 grams added. This pattern suggests that, for a 160 gram weight, the length of the spring may be extrapolated to be 50 + 3 = 53 centimetres. There can be no certainty about a value calculated by extrapolation.

See also **interpolation**.

LATIN *extra-*: out of, *polire*: to polish

Figure F1

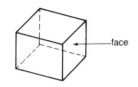

face Any of the flat sides of a **polyhedron**. Each face has straight **edges**, e.g. a cube has six faces, each of which has four edges.

factor
1 A whole number that exactly divides another whole number, e.g. 3 is a factor of 24 because $8 \times 3 = 24$. A prime factor is one that is a **prime number**, e.g. 2 and 3 are prime factors of 12, but the other factors of 12 (1,4,6,12) are not prime factors. A factor is also called a **submultiple**.
2 If **polynomials** are multiplied together, each is a factor of the product.
Example
$(a + b)$ and $(a - b)$ are factors of $a^2 - b^2$, because $(a + b)(a - b) = a^2 - b^2$
LATIN *factor*: maker

factorial [fac-*tor*-i-al] The function whose value is found by multiplying together all the positive whole numbers up to a given number.
   The symbol for factorial is !, so that 4! means $4 \times 3 \times 2 \times 1 = 24$, and 5! means $5 \times 4 \times 3 \times 2 \times 1 = 120$. n! is read 'n factorial'. It is sometimes written $\underline{|n}$. 0! is given the value 1.
   Factorials occur in the formulae for **binomial coefficients** and in the calculation of **permutations**.

factorise (v) To break up into factors.
Examples
a 114 may be factorised as $2 \times 3 \times 19$ or 6×19 or 3×38 or 2×57
b $x^2 + 3x + 2$ can be factorised into $(x + 1)(x + 2)$

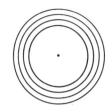

Figure F2

family A collection of related geometrical curves.
Example
Figure F2 shows a family of circles, related by having the same centre. Individual members of the family differ by having different values for the radius, r. In this case, r is known as the **parameter** for the family.

femto- Prefix meaning one thousand-million-millionth. Symbol f.
Example
1 femtogram = .000 000 000 000 001 grams
1 fg = 10^{-15} g
DANISH *femten*: fifteen

Fermat [fair-*ma*], **Pierre de** (1601–1665) A famous 17th century French mathematician who did much pioneering work in analytical geometry, **calculus**, the theory of numbers and probability theory. He was also interested in the applications of mathematics, particularly to optics.

Fibonacci [fee-bon-*ah*-chee], (?1170–1250), also known as Leonardo of Pisa (Italy)   In 1202, Fibonacci wrote his *Liber abaci* (Book of the abacus) in which he drew the attention of Europeans to the advantages over the Roman numeral system of the Hindu–Arabic numerals and methods of computation.

He is also remembered for the Fibonacci sequence of numbers: 0,1,1,2,3,5,8,13, ..., in which the next term is found by adding the two preceding terms. It is said that Fibonacci came across this sequence in a problem about rabbit reproduction. The sequence appears in many aspects of nature and art.
See also **golden mean**.

figurate numbers [*fig*-yoo-rayt] Sequences of whole numbers named after geometrical figures: triangle, square, pentagon, etc. They

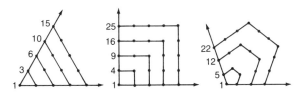

Figure F3

are formed as shown in Figure F3.
Also known as polygonal numbers.

figure 1 A geometrical shape. Circle, triangle, polygon are plane figures. Sphere, cube, polyhedron are solid figures.
2 A numerical symbol, such as a **digit** (0,1,2, ... 9).
LATIN *figura*: form or shape

finite (a) [*fy*-nyt]   Having a boundary or limit. Not **infinite**.
A finite **set** has a countable number of **elements**. The region inside a **polygon** is finite. The set of letters in the alphabet is finite. The set of **natural numbers** is not finite. A **line segment** has finite length, whereas a line is of infinite length.
LATIN *finitus*: bounded

Fisher, R. A. (1890–1962)   A British mathematician noted for the development of statistical methods to help in the design of experiments and the interpretation of experimental results. A technique known as analysis of **variance** has applications in many fields, including agriculture, psychology, biology, education and engineering.

flip   A term used in motion geometry to describe the result of **reflection**.
Example
In Figure F4, △DEF is the flip of △ABC after reflection in the line PQ.

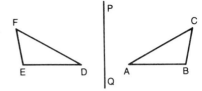

Figure F4

floating point Referring to the decimal point that, for ease of certain calculations, is allowed to 'float' away from its normal position separating the whole part from the fraction part of a number, e.g. the number 543.21 may be written as 5432.1×10^{-1}, 5.4321×10^2 etc.
See also **scientific notation**.

flow chart A diagram showing how to break down a problem or task into small steps. It consists of a sequence or network of boxes containing operating instructions and decision questions linked by arrows, which lead one from 'start' to 'stop'. Flow charts are often used in computer programming. Figure F5 shows the flow chart idea applied to a problem in everyday life.

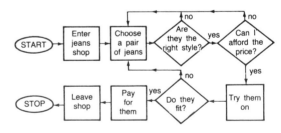

Figure F5

focus [*foh*-kus], *pl.*: **foci** [*foh*-sy] or **focuses** A special point related to a **conic section**.
Example
The points labelled F in Figure F6 are foci. P is a moving point in each case.
LATIN *focus*: a hearth

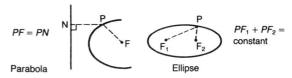

Figure F6

force In mechanics, an influence that can cause a change in the motion of an object, e.g. a push, a pull, the force of gravity, magnetic force.

The **standard unit** of force in the SI system is the newton (N). It is the force needed to accelerate a mass of 1 kilogram at the rate of 1 metre per second per second.

Force (F), mass (m) and acceleration (a), when measured in standard units, are related by the formula, $F = ma$.

formula [*form*-yoo-lah], *pl.*: **formulas** or **formulae** [*form*-yoo-lee]   An **equation** giving the relation between two or more quantities.
Example
The area (A square centimetres) of a triangle with base b centimetres and height h centimetres is given by the formula $A = \frac{1}{2}bh$.
LATIN *formula*: a little form

four-colour theorem   The theorem that, in colouring the different regions on a flat map, no more than four colours are needed if no adjacent regions are to have the same colour. This was first suggested by Francis Guthrie in 1852, but it was only in 1976 that it was proved to be true by Appel and Haken, using a computer.

fraction   Any **rational** number that is not an **integer**.
Examples
a   Proper fractions (numerator less than denominator): $\frac{1}{2}, \frac{3}{4}$
b   Improper fractions (numerator greater than denominator): $\frac{4}{3}, \frac{8}{5}$

c Decimal fractions: 0.1, 0.75
Fractions expressed as a ratio (e.g. $\frac{3}{4}$) are called common (or vulgar) fractions to distinguish them from decimal fractions.
Equivalent fractions have equal value, e.g. $\frac{4}{12} = \frac{3}{9} = \frac{2}{6} = \frac{1}{3}$
Any fraction may be expressed in percentage form, e.g. $\frac{3}{4} = 75\%$
Fractions occur in algebra as the ratio of one polynomial to another, e.g. $(ax + b)/(2ax^2 - b)$.
LATIN *fraction*: a break

frame of reference A fixed arrangement of points or lines used for describing the position of points or objects.

The most commonly used frames of reference in mathematics are for the **Cartesian** system of **coordinates**. These consist of two **axes** at right angles for describing the positions of points in a plane, and three axes at right angles for describing the positions of points in three-dimensional space.

frequency [*free*-kwen-see] In statistics, the number of items in a given category, or the number of times an event occurs.
Example
If 20 tosses of a coin result in 12 heads and 8 tails, the frequency of heads is 12, the frequency of tails is 8. The relative frequency of heads in this case is 12/20 or 0.6.
LATIN *frequentia*: a crowd

frequency distribution In statistics, a table listing a set of values of a variable together with the number of occurrences (the **frequency**) of each value.
Example
In a survey of 100 families, the variable of interest was the number of pets they kept. The following table shows the distribution of frequencies:

pets	0	1	2	3	4	5	6
families	10	40	29	9	8	4	0

A frequency distribution may be illustrated graphically by a **histogram**.
See also **cumulative frequency**.

frequency polygon   The graph of a **frequency distribution** produced by joining the midpoints of the tops of the columns of a **histogram**.
Example
Figure F7 shows the frequency polygon constructed from the data given in the table in the entry 'frequency distribution' above.

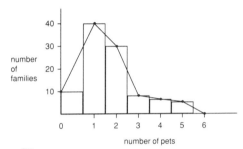

Figure F7

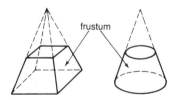

Figure F8

frustum   A solid formed from a **cone** or a **pyramid** by slicing through it along a plane parallel to the base (see Figure F8).
LATIN *frustum*: a piece

function   An important mathematical concept for describing the relation between two or more **variables**.
  Consider two variables, x and y. For y to be a function of x, the values of y must depend on the values of x, and in such a way that, for any particular value of x, there is only one possible value of y. x is known as the **independent** variable and y the **dependent** variable.
Example
The statement $y = x^2 + 1$ gives y as a function of x; the table below shows a selection of values of

function

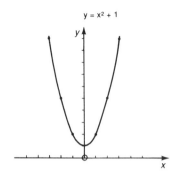

Figure F9

x and the corresponding values for y calculated from the above statement:

x	3	2	1	0	−1	−2	−3
y	10	5	2	1	2	5	10

Note that, although y has only one value for a given x, the same is not necessarily true in reverse. This is plain from the graph shown in Figure F9.

The idea of function is defined more precisely in **set** language. A function is a **mapping** of one set to another set so that to each element (x) of the first set (called the **domain**) there corresponds only one element (y) of the second set (called the **range**) (see Figure F10).

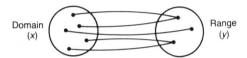

Figure F10

To denote a function in general, single letters are used, such as f, g, F, G. 'A function of x' is written $f(x)$ and read 'f of x'. In the above example, $y = f(x) = x^2 + 1$.

A function may have more than one independent variable, e.g. $y = f(x,z)$ may be $y = 2x + 3z$.

LATIN *functio*: performance, execution

G

Gauss [*gows*], **Carl Friedrich** (1777–1855) German mathematician noted for developments in number theory, algebra and geometry.
The **normal distribution** is sometimes called the Gaussian distribution in his honour. He used probability theory in the formulation of a pension plan and applied mathematics to the study of electricity, magnetism and gravity. From 1796 to 1814 he kept a diary, which has been published.

geo-board (abbreviation for geometry board) A flat board supporting a bed of nails arranged in regular rows and columns. Elastic bands can be stretched around the nails to form **polygons**. A geo-board is useful for studying the properties of geometrical shapes and their areas.

geodesic [jee-o-*dee*-sik] A path drawn on a curved surface and joining two points such that it has the shortest length. On a **sphere**, a geodesic is a path along a **great circle**. That is, it is an arc with its centre at the centre of the earth.
GREEK *geo*: earth, *daisia*: to divide

geometric mean Of two numbers, the square root of their product.
Example
The geometric mean of 3 and 12 is $\sqrt{(3 \times 12)} = 6$.
  Two numbers and their geometric mean form three terms of a **geometric progression**, as in the above example: 3,6,12.
Also called geometric average.

geometric progression A sequence of numbers in which each number (after the first) is found by multiplying the previous number by a fixed multiplier. This multiplier is called the common ratio.
Examples
a 1,2,4,8,16,...   (common ratio = 2)
b $1,\frac{1}{2},\frac{1}{4},\frac{1}{8},...$ (common ratio = $\frac{1}{2}$)
c 1,−3,9,−27,81,...   (common ratio = −3)
In general, a geometric progression can be represented by $a, ar, ar^2, ar^3, ...$ with the nth term = ar^{n-1}.
See also **arithmetic progression**.

geometric series The sum of the terms of a geometric progression.
Examples
a 1 + 2 + 4 + 8 + ...
b $1 + \frac{1}{2} + \frac{1}{4} + \frac{1}{8} + ...$
c 1 − 3 + 9 − 27 + ...
The sum of the first four terms of each of the above series is: **a** 15; **b** $1\frac{7}{8}$; **c** −20. In general, the sum of the first n terms is:

$$S_n = \frac{a(1 - r^n)}{1 - r}$$

where a is the first term and r is the common ratio.
A **recurring** decimal can be represented as a geometric series.
Example
$0.\dot{1} = 0.111... = 1/10 + 1/100 + 1/1000 + ...$, which is a geometric series with common ratio = 1/10.

geometry The major branch of mathematics that deals with the position, size and shape of figures in space.
  For over 2000 years, most developments in geometry accepted the work of **Euclid** (3rd century BC) as their foundation. One of the greatest of these developments occurred in the 17th century when **Descartes** (1596−1650)

applied **algebra** to the study of geometrical problems. A more modern development has been **topology**. The 19th and 20th centuries have seen the appearance of geometries that challenge the assumptions and **axioms** of **Euclidean** geometry, and these have had a profound effect on the understanding of space and of the logic of mathematical thinking.
GREEK *geometria*: land measuring

giga- [*gee*-ga]   A prefix meaning one thousand million times. Symbol G.
Example
   1 gigatonne = 1 000 000 000 tonnes
       1 Gt = 10^9 t
GREEK *gigas*: a giant

given (a)   To be accepted as already known.
Examples
a   Given that $a = 10$, then $2a = 20$
b   Given that triangle ABC is equilateral, it follows that each of its angles is 60°
Given is sometimes used as a noun, e.g. in example **a** above, '$a = 10$' is a given.

glide   A term used in motion geometry to describe the effect of a **translation**.
Example
The movement of $\triangle$ABC to a new position DEF without **rotation** or **reflection** is a glide.

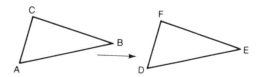

Figure G1

Gödel [*ger*-dl] (1906–1978)   A prominent American mathematician whose work has led to a fundamental reassessment of the logical foundations of mathematics. A Gödel statement is one that claims it cannot be proved.

golden ratio (or **section** or **mean**) A special way of dividing a line segment into two parts so that the ratio of the smaller part to the larger is the same as the ratio of the larger to the whole. If the two parts have lengths a and b, then $a/b = b/(a + b)$. This ratio can be shown to equal $2/(1 + \sqrt{5})$ or 0.618 approximately.

A rectangle with sides in the ratio 1:0.618 is called a golden rectangle. Because it is pleasing to the eye, it has been much favoured by some artists and designers.

googol The number 10^{100}, that is, 1 followed by one hundred zeros. It is probably larger than the number of protons and neutrons in the universe.

gradient [*gray*-di-ent] The slope or steepness of a line or curve. A way of measuring gradient is by reference to **Cartesian coordinates** as in Figures G2 and G3.

1 For a straight line, as in Figure G2, P,Q, are any two points on the line. The gradient of the line is measured by the ratio,

$$\frac{\text{increase in } y \text{ when moving from P to Q}}{\text{increase in } x \text{ when moving from P to Q}}$$

i.e., gradient = QN/PN. This is the **tangent** of the angle, θ, that the line makes with the x-axis: gradient = $\tan \theta$. The line in Figure G2 has a positive gradient. For a line that has a downhill slope from left to right, the gradient is negative. A horizontal line has zero gradient.

2 For a curve, the gradient changes from point to point. At a given point, P, in Figure G3 the gradient is found by drawing a **tangent** to the curve at that point and then measuring the gradient of the tangent as in the first example. Again, gradient = $\tan \theta$.

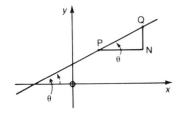

Figure G2

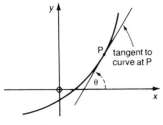

Figure G3

Gradient represents the rate of change of the **function** that is represented by the line or curve. This is an important idea in **calculus**.

LATIN *gradus*: a step

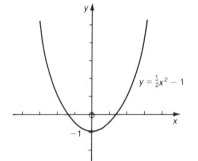

Figure G4

graph A diagram or picture designed to show the relation between two or more quantities. Some examples are: **bar graph, broken line graph, histogram, circle graph, scattergram.**

Graphs play an important part in coordinate geometry, where they are used to study the properties of **functions**. In Figure G4, the graph of the function $y = \frac{1}{2}x^2 - 1$ illustrates a number of properties of the function: its **symmetry**, its **minimum** value, the values of x for $y = 0$, etc.

GREEK *graphe*: drawing, writing

graph paper Paper on which intersecting lines are drawn to assist in the drawing of graphs. A common type has lines parallel to the edges of the paper and drawn at millimetre intervals. This enables graphs to be drawn with **Cartesian** axes. Some graph paper has one or both sets of lines spaced at logarithmic intervals. It is then called log paper.

gravity In science, the **force** of attraction exerted on one object by another. This force is regarded as existing between any two objects anywhere in the universe, but it is most commonly applied to the force exerted by the earth on an object near the earth's surface. This force then appears as the weight of the object.

great circle A circle drawn on a sphere and having the same centre as the sphere.

There is only one great circle that passes through two points on the surface of a sphere, and this circle provides the shortest path between the two points.

Any other circle drawn on the surface of a sphere has a smaller radius than a great circle.

If the earth is regarded as a sphere, then the equator and every **meridian** of **longitude** are great circles.
See also **geodesic, small circle**.

grid A network of parallel lines, often like **graph paper**, drawn on a map to help locate positions on the map.

greatest common factor (GCF) The largest number that is a **factor** of every number in a set.
Example
Consider the **prime** factors of 12, 36, 60:
$$12 = 2 \times 2 \times 3$$
$$36 = 2 \times 2 \times 3 \times 3$$
$$60 = 2 \times 2 \times 3 \times 5$$
From here it can be seen that 2 and 3 are common prime factors and that the greatest common factor is $2 \times 2 \times 3 = 12$. Greatest common factors occur also in algebra, e.g. the GCF of $3a(a + 2)$ and $6(a + 2)^2$ is $3(a + 2)$.

grouping A method of arranging the terms of an algebraic expression to make it easier to **factorise** the expression.
Example
$x^2 + xy - 2x - 2y$. Group the terms in pairs so that the pairs have a **common factor:**
$x(x - y) - 2(x + y)$. The factors of the original expression are then: $(x + y)(x - 2)$.

H

Figure H1

half line The part of a **line** on one side of a given point. Also called a **ray** (see Figure H1).

harmonic mean The reciprocal of the **arithmetic mean** of the reciprocals of a set of positive numbers.
Example
To find the harmonic mean of 3 and 6: the reciprocals are 1/3 and 1/6 and these have mean 1/4. Therefore the harmonic mean of 3 and 6 is 4. Note that 3,4,6 are terms of a **harmonic progression** (because their reciprocals 1/3, 1/4, 1/6 or 4/12, 3/12, 2/12 are in **arithmetic progression**).

harmonic progression A sequence of numbers whose **reciprocals** form an **arithmetic progression**, e.g. $1, \frac{1}{2}, \frac{1}{3}, \frac{1}{4}, \ldots$

Similar musical strings with the same tension sound in harmony with each other if their lengths are proportional to the terms of a harmonic progression.
GREEK *harmonia*: fitting together

hectare [*hek*-tair] Abbreviation: ha. A non-SI unit of **area**, used in land measurement.
$$1 \text{ hectare} = 10\,000 \text{ square metres}$$
$$1 \text{ ha} = 10^4 \text{ m}^2$$

hecto- A prefix meaning one hundred times. Symbol h.
Example
$$1 \text{ hectolitre} = 100 \text{ litres}$$
$$1 \text{ hL} = 10^2 \text{ L}$$
GREEK *hekaton*: hundred

height **1** The length of an **altitude** of a geometrical figure. **2** The distance of a point vertically above a reference line or plane, as in 'height above sea-level'.

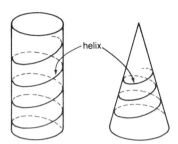

Figure H2

helix [*hee*-liks] The curve that results from drawing a straight line on a sheet of paper and then wrapping it around a **cylinder** or **cone**. (see Figure H2).
Examples
a A screw thread is an example of a helix.
b The DNA molecule follows the shape of a double helix.
GREEK a spiral

hemisphere [*hem*-i-sfeer] Half a **sphere**, formed by making a plane cut through the centre of the sphere.
GREEK *hemi*: half

heptagon A seven-sided **polygon**.
GREEK *hepto-*: seven, *gonia*: angle

hexadecimal Of a number system based on sixteen. Abbreviation: hex.
  In the hexadecimal system of notation, the following **digits** are used: 0 1 2 3 4 5 6 7 8 9 A B C D E F, and the number sixteen is written '10'. Hexadecimal notation is used in computing.
Example
The hex numeral F7 converted to decimal notation is $15 \times 16 + 7 = 247$, and the hex numeral BD is $11 \times 16 + 13 = 189$.
GREEK *hexa-*: six and LATIN *decem*: ten

hexagon A six-sided **polygon**. A regular hexagon may be subdivided into **equilateral** triangles by joining the **vertices** to the centre. Regular hexagons fit together to form a **tessellation** (see Figure H3).
GREEK *hexa–*: six, *gonia*: angle

Figure H3

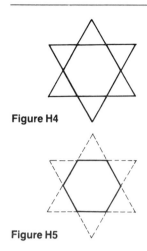

Figure H4

Figure H5

hexagonal (a) [heks-*ag*-o-nal] Having the shape of a **hexagon**.

hexagram A star-shaped figure made from two intersecting equilateral triangles, as shown in Figure H4. It may also be made by extending the sides of a regular **hexagon** until they intersect, as in Figure H5.
GREEK *hexa-*: six, *-gramma*: drawing

hexahedron A solid figure with six plane faces. A regular hexahedron is a **cube** and is one of the five possible shapes for a **regular polyhedron**.
GREEK *hexa-*: six, *hedra*: base

hexomino [heks-*om*-in-oh] A plane figure made from six equal squares joined so that some sides are shared.
Examples can be seen in Figure H6.

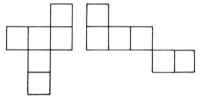

Figure H6

highest common factor (HCF) The largest number that is a **factor** of every number in a set.
Example
Consider the **prime** factors of 12, 36, 60:
$$12 = 2 \times 2 \times 3$$
$$36 = 2 \times 2 \times 3 \times 3$$
$$60 = 2 \times 2 \times 3 \times 5$$
From here it can be seen that 2 and 3 are common prime factors and that the highest common factor is $2 \times 2 \times 3 = 12$. Highest common factors occur also in algebra, e.g. the HCF of $3a(a + 2)$ and $6(a + 2)^2$ is $3(a + 2)$.

Hindu–Arabic numerals See **Arabic numerals**. The history of these numerals can be traced to the Hindu civilisation in India over 1200 years ago. Their modern form was developed by Arabian mathematicans in the Middle Ages.

histogram Graph of a **frequency distribution** in which equal intervals of values are marked on a horizontal axis and the frequencies associated with these intervals are indicated by the areas of rectangles erected vertically on these intervals.
Example
The histogram in Figure H7 shows the number of students achieving different scores in an English test taken by a group of fifty students.
GREEK   *histos*: web or tissue, -*gramma*: drawing

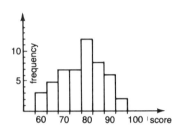
Figure H7

homogeneous (a) [hoh-moh-*jeen*-i-us]
Describing a **polynomial**, all terms of which are of the same **degree**.
Example
$2x^3 - 5x^2 y + 4xy^2 - y^3$ is a homogeneous polynomial of the third degree. (*Compare* $2x^3 - 5x^2 + 1$, which is a polynomial of the third degree, but is not homogeneous.)
GREEK   *homo*: same, *genes*: kind

horizontal (a)   Parallel to the horizon; level; at right angles to the **vertical** direction.
  Every line drawn on a horizontal **plane** is horizontal.
  The *x*-axis on a **Cartesian** graph is often called the horizontal axis.
GREEK   *horizein*: to act as boundary

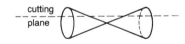
Figure H8

hyperbola [hy-*per*-bo-la]   One of the **conic sections**. It is formed by a plane cutting through both bases of a double **cone** (Figure H8), producing a plane curve with two parts (arms) (Figure H9).
  The hyperbola may also be considered as the

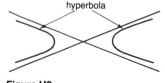
Figure H9

hypocycloid

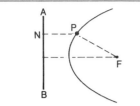

Figure H10

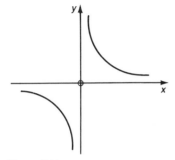

Figure H11

locus of a point (P) (Figure H10) such that the ratio of the distance of P from a focal point (F) to its distance from a fixed line (AB) remains constant and greater than 1. In Figure H10, P will trace out a hyperbola if PF is greater than PN and if PF/PN remains constant.

In algebra, if y varies inversely as x, then $y = k/x$ where k is a constant and the graph of y against x is a hyperbola (Figure H11). This is an example of a rectangular hyperbola, and the x- and y- axes are **asymptotes** to the curve.

GREEK *hyperbole*: throwing beyond

hyperboloid [hy-*per*-bo-loyd]   A solid formed by rotating both arms of a **hyperbola** about one of the axes of symmetry.

If AB is the axis of rotation in Figure H12, one solid (something like a pulley wheel) is the result (see Figure H13). If CD is the axis of rotation, two cup-shaped solids are formed, as shown in Figure H14.

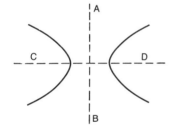

Figure H12

Figure H13

Figure H14

hyperspace   A space in which more than three **coordinates** are needed to describe the position of a point. Sometimes the combination of ordinary three-dimensional space with the time dimension is referred to as a hyperspace.

GREEK *hyper*: beyond

hypocycloid   A curve traced out by a point on the circumference of a circle as it rolls around the inside of a larger circle in the same plane,

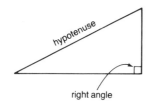

Figure H15

e.g. an **astroid** is a hypocycloid.
GREEK *hypo*: under, *kuklos*: circle

hypotenuse [hy-*pot*-e-nyooz]   In a right-angled triangle, the side opposite the right angle (see Figure H15). It is the longest side.
GREEK *hypo*: under, *teinein*: to stretch

hypothesis [hy-*poth*-e-sis]   **1**   A statement that seems to explain some observations, but which has as yet to be proved.   **2**   (statistics) — As in 'null hypothesis'. When aiming to discover if observations about two groups reveal any basic difference between the groups, the null hypothesis, that there is no difference, is first made.
Example
In investigating the mathematical ability of girls and boys, the hypothesis is made that there is, in general, no difference. It then remains to be seen whether observations made on samples of boys and girls disprove this hypothesis.
GREEK *hypothesis*: foundation

I

i The symbol used to represent the number $\sqrt{-1}$.
In elementary arithmetic, the square root of a negative number has no meaning, but in more advanced mathematics, $\sqrt{-1} = i$ is introduced as a solution to the equation $x^2 = -1$. All other negative numbers may then have a square root, e.g. $\sqrt{-16} = 4i$.
Euler introduced the symbol, i, about the year 1750. Numbers expressed with i are called **imaginary** numbers.
See also **real numbers**.

icosahedron A solid figure with twenty plane faces. It is possible to construct a regular icosahedron; its faces are equilateral triangles, and it is one of the five possible shapes for a **regular** polyhedron.
GREEK   *eikosi*: twenty, *hedra*: base

identity 1 A mathematical statement that is true for all values of the variables.
Example
These two statements are identities because they are true for every value of x or A:
$2(x + 1) = 2x + 2$,
$\tan A = \sin A/\cos A$.
The equation $2(x + 1) = 4$ is not an identity; it is true if $x = 1$, but not for any other value of x. To distinguish between the two kinds of statement, the sign $\equiv$ is sometimes used instead of $=$ for an identity: $\tan A \equiv \sin A/\cos A$.
2 In the language of set theory, an element that does not change the value of another element during an operation.

Example
0 is the identity element for addition and 1 is the identity element for multiplication, because $a + 0 = a$, and $a \times 1 = a$.
LATIN *idem*: the same

iff The abbreviation for 'if and only if'. This phrase is used to connect two statements when the truth of each one is conditional on the truth of the other.
Example
It is true that, if the three angles of a triangle are equal, then the three sides are equal. It is also true that, if the three sides of a triangle are equal, then the three angles are equal. More concisely, the three angles of a triangle are equal iff the three sides are equal.

image
1 A point, line or figure after **reflection** or other transformation.
Example
$\triangle A'B'C'$ in Figure I1 is the image of $\triangle ABC$ as reflected in the line.

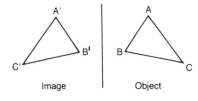

Figure I 1

2 A set of numbers related to a given set by a **function** rule.
Example
If $y = x^2$, then the values 1, 2, 3 for x produce values 1, 4, 9 for y. In this case, the set $\{1,4,9\}$ is the image of set $\{1,2,3\}$.
LATIN *imago*: an imitation

imaginary number A number with a negative square.
Example
$\sqrt{-16}$ is an imaginary number because its square is -16. $\sqrt{-16}$ is usually written $i\sqrt{16}$ or $4i$ with i standing for $\sqrt{-1}$.

Imaginary numbers are not used in elementary arithmetic, but their use in higher mathematics allows many important

applications, such as in the study of electric circuits.
See also **i, number.**

implicit (a) [im-*plis*-it] Not stated outright. An implicit **function** is one in which the dependent variable is not stated directly as a function of the independent variable.
Example
$xy = 12$ gives y as an implicit function of x. Compare with $y = 12/x$, in which y is stated explicitly as a function of x.
LATIN *implicare*: to enfold

improper fraction A **fraction** that is expressed with a numerator and a denominator, the numerator being greater than the denominator, e.g. 5/2.
An improper fraction can be expressed as a mixed number, e.g. $5/2 = 2\frac{1}{2}$. An improper fraction always has a value greater than 1.
Compare **proper fraction.**

incentre The centre of a circle drawn inside a triangle or polygon so as to touch every side. The incentre of a triangle is the point where the angle bisectors meet.
See also **incircle.**

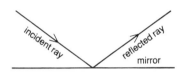

Figure I 2

incident (a) [*in*-sid-ent] Falling on or meeting, as a ray of light falling on a surface and then being reflected (see Figure I2).
LATIN *incidens*: befalling

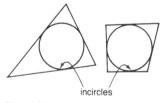

Figure I 3

incircle A circle drawn inside a triangle or polygon so as to touch every side (see Figure I3).
Incircles are possible for all triangles, for all regular polygons and for some non-regular polygons (Figure I3). Many non-regular polygons cannot have an incircle (see Figure I4).
See also **incentre.**

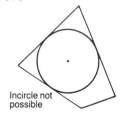

Figure I 4

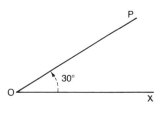

Figure I 5

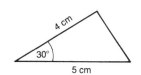

Figure I 6

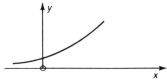

Figure I 7 Graph of an increasing function

inclination The angle between two lines, or between two planes, or between a line and a plane.
Example
In Figure I5, the inclination of line OP to line OX is 30°.
  The inclination of the earth's axis to the plane of the earth's orbit is $23\frac{1}{2}°$ approximately.
LATIN *inclination*: leaning

included angle The angle between two sides of a triangle, the lengths of which are known.
Example
In Figure I6, the angle marked is the included angle. The other two angles are not included angles.
  If two side lengths of a triangle and the size of the included angle are known, there is enough information to construct the triangle.

increasing (a)   Growing greater.
Example
In Figure I7, y is said to be an increasing function of x if y continuously grows greater as x becomes larger.
LATIN *increscere*: to grow

increment [*in*-krem-ent] The increase as a variable changes from one value to another. This increment may be positive, zero or negative. The symbol for increment is the Greek letter delta, δ or $\triangle$.
Examples

Values of x		Increment
1st val. (x_1)	2nd val. (x_2)	$\delta x = x_2 - x_1$
8.5	8.7	+0.2
0.65	0.65	0
141	138	−3

If y is a function of x, there will be corresponding increments for y as x changes. The study of small increments becomes important in **calculus**.
LATIN *incrementum*: an increase

independent (a) Not relying on another.
1 Independent variable. A variable whose values may be freely chosen and upon which the values of other variables depend.
Example
In $y = 2x + 1$, values of x may be chosen independently, and then values of y follow according to rule. x is the independent variable and y the **dependent** variable.

In graphing, it is usual to plot the independent variable along the horizontal axis and the dependent variable along the vertical axis.
2 Independent event. Two events are independent of each other if the **probability** of one happening is not affected by whether the other happens.
Example
In tossing a coin, the probability of a head is 1 in 2, or $\frac{1}{2}$. Whether one toss produces in fact a head or a tail has no effect on the result of the next toss. The two tosses are independent events. To calculate the probability of two independent events producing a particular result, the individual probabilities are multiplied together. In the example, the probability of getting 2 heads with 2 tosses is $\frac{1}{2} \times \frac{1}{2} = \frac{1}{4}$.
LATIN   *in-*: not, *dependere*: to hang upon

indeterminate (a) [in-de-*term*-in-ayt]   Not able to be given a value.
Example
The expression $3(x - 2)/(x - 2)$ has the value 3 for all values of x except $x = 2$. Substituting $x = 2$ produces the result 0/0, a symbol that has no meaning in arithmetic; so the expression is indeterminate for $x = 2$.
Calculus provides ways of dealing with problems like this one.
LATIN   *in-*: not, *determinare*: to limit

index (*pl*.: **indices** [*in*-di-seez])   A number expressing a **power**.

Example
In $y = 4x^2$, 2 is an index meaning that x is raised to the power of 2.
Also called an **exponent**.
LATIN *index*: forefinger, sign

index laws Rules for working with indices or powers.

Rule	Example
$a^m \times a^n = a^{m+n}$	$x^2 \times x^3 = x^5$
$a^m \div a^n = a^{m-n}$	$x^5 \div x^2 = x^3$
$(a^m)^n = a^{mn}$	$(x^2)^3 = x^6$

These rules follow from the definition of a^n where n is a whole number: $a^n = a \times a \times a \times \ldots$ n times. The rules are then extended for non-integral values of n, e.g. $a^0 = 1$; $a^{\frac{1}{2}} = \sqrt{a}$.

indivisible (a) Of a whole number, not able to be exactly divided. A prime number is indivisible: it has no factors (other than itself and 1). Other whole numbers may be indivisible by a particular number, e.g. 10 is indivisible by 3.

induction
1 A method of reasoning by drawing a general conclusion from a number of individual facts. Such reasoning provides good support for accepting the conclusion, but cannot prove it beyond all doubt.
Example
If the only swans you ever saw were white, it would be reasonable to conclude that all swans were white. One black swan would, however, disprove your conclusion.
Compare **deduction**.
2 Mathematical induction is the name for a different form of reasoning. Properly carried out, it must lead to a true conclusion.
   It is applied to certain theorems about **integers** and follows these steps:
(1) Show that if the theorem is true for $n = r$, then it is also true for $n = r + 1$.

(2) Show that the theorem is true for $n = 1$.
(3) Then it follows that the theorem is true for $n = 2,3,4, \ldots$ and all values of n, where n is an integer.

LATIN *in-*: in, *ducere*: to lead

inequality   A statement that one quantity is not equal to another. An inequality may be expressed in the following ways:

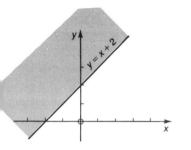

Figure I 8

$a \neq b$ 'a is not equal to b'
$a < b$ 'a is less than b'
$a \leq b$ 'a is less than or equal to b'
$a > b$ 'a is greater than b'
$a \geq b$ 'a is greater than or equal to b'
Example
The inequality $y \geq x + 2$ is represented by the unshaded region on the graph, shown in Figure 18.

inequation   A statement of **inequality** involving an unknown for which values can be calculated. The **solution** of an inequation can be compared with the solution of an **equation**.
Example

Equation	Inequation
$2x + 6 = 30$	$2x + 6 < 30$
$2x = 24$	$2x < 24$
$x = 12$	$x < 12$

inference   **1**   Any method of reasoning from data to conclusion, e.g. if two angles of a triangle are 30° and 60°, then by inference the third angle is 90°.   **2**   The conclusion itself, e.g. in the above example, the inference is that the third angle is 90°.

In mathematics, the main valid methods of arriving at an inference are through **deduction** and mathematical **induction**.

LATIN   *in-*: in, *ferre*: to carry

infinite (a) [*in*-fin-it]   Having no boundary or limit. Not **finite**.

infinitesimal   114

Example
The **sequence** of whole numbers, 1,2,3, ... is an example of an infinite sequence. It has no end.
    An infinite **set** has the property that its elements can be put into one-to-one correspondence with the elements of one of its subsets,
Example
The set of whole numbers bears a one-to-one relation with the even numbers:

$$\begin{array}{ccccc} 1 & 2 & 3 & 4 & 5 \ldots \\ \updownarrow & \updownarrow & \updownarrow & \updownarrow & \updownarrow \\ 2 & 4 & 6 & 8 & 10 \ldots \end{array}$$

LATIN   *in-*: not, *finitus*: bounded

infinitesimal (a)   Infinitely small; approaching zero.
    Infinitesimal **calculus** is a branch of mathematics that approaches problems about rates of change and about areas and volumes from the idea of limiting values, using very small **increments**.
LATIN   *in-*: not, *finitus*: bounded

infinity [in-*fin*-i-tee]   Infinity is a word expressing the idea that for any number, no matter how great, there is a greater number, and that for any point, no matter how distant, there is a further point. The sequence 1,2,3, ... has no end and is said to continue to infinity. A line stretches unendingly in space to infinity. Infinity is not a definite number or definite position. It is represented by the symbol ∞.
Example of the use of the symbol ∞:
As $n \to \infty$, $1/n \to 0$. This statement means 'as n increases without limit, $1/n$ becomes smaller and smaller, approaching the value 0'.
See also   **limit**.
LATIN   *in-*: not, *finitus*: bounded

inflection (or **inflexion**)   The change of curvature of a graph or surface from **concave** to

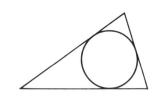

Figure I 9

Figure I 10

convex or vice versa.
See **point of inflection**.
LATIN *inflectere*: to bend

inscribe (v) To construct one figure inside another so that the two figures have common points but do not intersect.
Example
Figure I9 shows a circle inscribed in a triangle, and Figure I10 a hexagon inscribed in a circle.
See also **incircle, circumcircle**.
LATIN *in-*: in, *scribere*: write

integer [*in*-t-jer] Any of the numbers:
$\ldots -3, -2, -1, 0, 1, 2, 3, \ldots$
$1, 2, 3, \ldots$ are the positive integers, whereas $-1, -2, -3, \ldots$ are the negative integers.
Compare **natural number, whole number** and *see also* **number**.
LATIN *integer*: whole

integral 1 Adjective for **integer**, e.g. integral values of a variable, x, include values like $3, 7, -2$, but exclude values like $0.5, 7.9$. The integral part of a number is the part that is an integer, such as 3 in $3\frac{1}{2}$ and 3.42. 2 An important type of function in **calculus**, used in the calculation of areas and volumes. $\int$ is the integral sign.

intercept The point on one of the axes of a graph where the graph cuts.
Example
In Figure I11, the x-intercept is -2 and the y-intercept is $+3$.
LATIN *intercipere*: to seize in passing

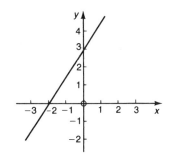

Figure I 11

interest Payment charged to a borrower by the lender of money or goods, or payment made to a lender by the borrower of money or goods. It is usually expressed as a certain percentage of the value of the loan payable per day, month or year.

interior 116

Example
A rate of 15% per annum means $15 is charged every year for every $100 lent.
See also **compound interest, simple interest**.
LATIN *interest*: is of importance

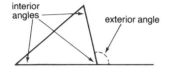
Figure I 12

interior (a) Inside. Opposite of **exterior**. The inside angles of a triangle or polygon are called interior angles when it is necessary to distinguish them from exterior angles (see Figure I12).
See also **co-interior angles**.

interpolate (v) [in-*ter*-po-layt] To calculate or estimate a value between two others, already known.
Example
Knowing that the temperature one day was 23°C at 2.00 p.m. and 24°C at 2.30 p.m., a value may be interpolated to estimate the temperature at 2.15 p.m. as 23.5°C.
See also **extrapolation**.
LATIN *inter-*: between, *polire*: to polish

interquartile range In statistics, the difference between the lower and upper **quartiles** of a **frequency distribution**. This is the section of the range of a variable associated with the middle 50% of the population under consideration.
Example
If the middle 50% of the workers in a large factory earn annual wages between $18 000 and $35 000, then the interquartile wages range is $35 000 − $18 000 = $17 000. In another large factory, the interquartile range may be $15 000, and so a comparison can be made about the **spread** of wages in the two factories. This often gives a better overall comparison than the full **range**, which would be affected by just one or a few very high or very low values.

intersect (v) **1** (geometry) — To have one or more points in common. **2** (set theory) — To have one or more elements in common.
See **intersection**.
LATIN *inter-*: between, *secare*: to cut

intersection
1 (geometry) — One or more points held in common by two or more figures.
Example
In Figure I13, points P and Q are the intersection of the line and the circle.
2 (set theory) — The set of one or more elements held in common by two or more sets.
Example
The intersection of set A = {1,2,4,8} and set B = {0,1,2,3} is the set {1,2}. This is written A ∩ B = {1,2} and read 'A cap B'. If two sets have no elements in common, it is said that their intersection is the **empty set**.
e.g. {A,B,C} ∩ {1,2,3,4} = ∅.
Two intersecting sets may be represented diagrammatically by two intersecting circles, with their common region shaded to represent the intersection of the sets, as in Figure I14.
See also **disjoint**.
LATIN *inter-*: between, *secare*: to cut

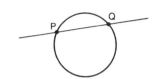
Figure I 13

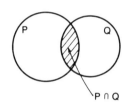
Figure I 14

interval 1 Lapse of time between two moments, e.g. between 3 p.m. and 4 p.m. there is an interval of 1 hour. **2** The distance or difference between two numbers or quantities.
Example
Contour lines on a map may be drawn at intervals of 20 metres, meaning that the space between one line and the next represents a difference in height of 20 metres.
3 The set of numbers (or points) between two fixed numbers (or points). If the end numbers (or points) are included in the set, the interval is

closed. If they are not included, the interval is open. (Closed interval shown as ●—●, open ○—○).
Example
Consider the interval from 2 to 3 on the number line:

and let n be any number in that interval. Then, if the interval is closed, the numbers 2 and 3 are included, so that $2 \leq n \leq 3$.
If the interval is open, 2 and 3 are not included, so that $2 < n < 3$.
In both cases, 2 and 3 are the end-points of the interval.
LATIN *intervallum*: space between walls

inverse The inverse of an **operation** reverses its effect. The inverse of multiplication is division.
Example
10 multiplied by 3 and then divided by 3 becomes 10 again. Addition and subtraction are the inverse of each other. Squaring and taking the square root are also inverses.
Also known as inverse operation.
LATIN *invertere*: to turn about

inverse function The function obtained when the dependent variable and independent variable of certain functions are interchanged.
Example
If $y = \log_{10}x$, then $x = 10^y$ is the inverse function. If $x = \tan \theta$, then $\theta = \arctan x$: arc tan x is the inverse tangent function, also written as $\tan^{-1}x$.
In general, the inverse function to function $f(x)$ is written $f^{-1}(x)$.

inverse proportion The relation between two variables such that, as the value of one variable increases, the value of the other

isogon

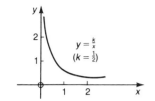

Figure I 15

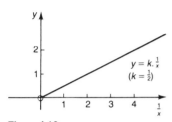

Figure I 16

variable decreases, the product of the two values remaining constant. If the two variables are x and y, then xy remains constant. The relation is written in one of the following ways:

$$xy = k, \text{ where k is a constant}$$
$$y = k/x$$
$$y \propto 1/x$$

and is read 'y varies inversely as x' or 'y is inversely proportional to x'.

If the values of y are plotted against values of x, the graph is a **hyperbola** (Figure I15). If values of y are plotted against values of $1/x$, the graph is a line through the origin (Figure I16). Also called inverse variation.

inverse square law A relation stating that an effect decreases in proportion to the square of the distance from the source of the effect. Such a relation occurs in the study of light, gravitation, magnetism, sound and other parts of science. If y represents the effect (for example light intensity) and x is the distance from the source, then the relation is expressed mathematically as $y \propto 1/x^2$ or $y = k/x^2$.

involution In algebra, the operation of raising a number to a power. Squaring and cubing a number are examples of involution. The **inverse** operation is called **evolution**.

irrational (a) Not **rational**. Describing a number that cannot be expressed as the **ratio** of two **integers**, e.g. $\sqrt{2}$, π, e, $\log_{10}3$.

In the decimal system, an irrational number appears as a non-repeating infinite decimal, e.g. $\pi = 3.141592\ldots$

isogon [y-so-gon] 1 A polygon with all angles equal, as in Figure I17. 2 A line on a map joining places having the same angle difference between magnetic north and true north.
GREEK *isos*: equal, *gonia*: angle

Figure I 17

isometric (a) Describing any **transformation** of a geometric figure that leaves the distance between any two points in the figure unchanged. **Reflection, rotation** and **translation** are isometric transformations. **Dilatation** is not isometric.
GREEK *isos*: equal, *metron*: measure

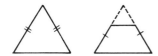

Figure I 18

isosceles (a) [y-*sos*-e-leez] Having two sides equal.
Examples
Isosceles triangle, isosceles trapezoid (see Figure I18)
GREEK *isos*: equal, *skelos*: leg

iteration [it-er-*ay*-shun] A method of calculation and of solving equations by making successive approximations until the desired degree of accuracy is reached. Iterative methods are particularly well suited to computer use.
LATIN *iteratio*: a repetition

J

joint variation In algebra, a type of **variation** in which the value of one variable depends on the values of two or more variables.
Example
The distance a car travels depends on both its speed and the length of time it travels. If the speed is uniform, this can be expressed as a formula, $d = kvt$, where d is the distance, v is the speed and t the time, with k being a constant depending on the choice of units.

K

kilo- [*kil*-oh]   A prefix meaning one thousand times. Symbol k.
Example
$$1 \text{ kilometre} = 1000 \text{ metres}$$
$$1 \text{ km} = 10^3 \text{m}$$
GREEK   *chilioi*: one thousand

kinematics [kyn-e-*mat*-iks]   The study of the motion of objects without reference to their mass or to the forces acting on them. Problems in kinematics relate to quantities such as displacement, velocity and acceleration.
See also   **dynamics, statics**.
GREEK   *kinema*: motion

Figure K1

kite   A **convex quadrilateral** with two pairs of equal **adjacent** sides (see Figure K1).
A non-convex quadrilateral with two pairs of equal adjacent sides is a chevron (see Figure K2).

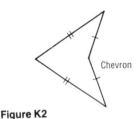

Figure K2

Königsberg bridge problem [*ke(r)n*-igs-berg]   A famous mathematical problem relating to the seven bridges over the river Pregel at the town of Königsberg in Germany. *Problem*: Can a person start at any point in Figure K3, walk over every bridge once only and so return to the starting point? **Euler** proved this to be impossible, using a branch of mathematics known as **topology**.

Figure K3

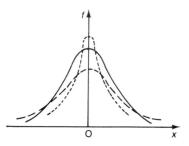

Figure K4

kurtosis [ker-*toh*-sis]   A measure of the shape of a frequency distribution as used in the study of statistics. Figure K4 shows three examples of a frequency curve, all with the same mean but each with a different kurtosis.

GREEK   *kyrtosis*: curvature

latitude One of the **coordinates** used for locating the position of a place on the earth's surface. (The other is **longitude**.) It is the angular distance of the place north or south from the equator.

The latitude of the equator is 0°, the latitude of the north pole is 90°N, the latitude of the south pole is 90°S. The latitude of the point P in Figure L1 is 30°S. All places with the same latitude lie on a parallel of latitude, which is a circle on the earth's surface parallel to the equator.

LATIN *latitudo*: breadth

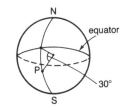

Figure L1

law of large numbers In statistics, the observation that, for a large number of independent repetitions of an experiment, the frequencies of the various possible outcomes tend to be in proportion to the **probabilities** of those outcomes.

Example
Consider the experiment of tossing two coins. There are four possible outcomes for each throw (HH, HT, TH, TT) so that the probability of obtaining two heads is 1/4. The law predicts that, for a large number of throws, about one-quarter of them will result in two heads, and that the greater the number of throws, the more certain one is that the proportion will approximate to one-quarter.

This is different from the so-called 'Law of averages', which would have us believe (falsely) that the more often the result is not two heads, then the better the chance that the next throw will result in two heads.

least common denominator (LCD)
A special application of **least common multiple**, used when the numbers concerned are **denominators** of fractions.
Example
To find the sum $\frac{1}{2} + \frac{1}{3} + \frac{1}{8}$, it is useful to convert the fractions so that they share the least common denominator:

$$\frac{12}{24} + \frac{8}{24} + \frac{3}{24} = \frac{23}{24}$$

least common multiple (LCM)
The smallest number that is a **multiple** of several other numbers.
Example
The least common multiple of 2,6,8 is 24. (2,6,8 have many common multiples: 24,48,72,96,..., but 24 is the least of them.)

Leibniz [*lyb*-nits], Gottfried Wilhelm (1646–1716)
A distinguished German scholar famous in many fields, including mathematics, philosophy, history and law. In mathematics, he is best known for having developed the **calculus**. He did this at the same time as **Newton**, although the two worked independently.

length
One-dimensional extent measured in units defined by a line segment. The **SI** standard unit of length is the metre.

To apply the notion of length to a curve, the curve may be approximated as a succession of very small line segments, and then the limit of their sum calculated as each segment approaches zero length (see Figure L2). Such a calculation can be accurately made using **integral calculus**.

Figure L2

leptokurtic (a)
In statistics, referring to a distribution that is more concentrated around the mean than a **normal distribution**. The graph

has a sharper peak than a **normal curve**.
Contrast **platykurtic**.
GREEK *lepto-*: thin, *kyrtosis*: curvature

light-year A unit of distance used in astronomy. It is the distance travelled by a pulse of light in 1 year. The speed of light is constant at approximately 3×10^8 metres per second. It follows that 1 light-year is approximately 9.46×10^{15} metres.

The nearest star to our sun is Alpha Centauri, which is distant 4.3 light-years.

like terms In algebra, **terms** that are the same, apart from their numerical **coefficients**.
Example
$3ax^2$ and $17ax^2$; $4x^2y$ and $11x^2y$. These are pairs of like terms, but $4x^2y$ and $11xy^2$ are not like terms.

Like terms can be added by adding the numerical coefficients, e.g. in the first two examples above, the sums may be written simply as $20ax^2$ and $15x^2y$; but, in the third example, the terms cannot be combined in the same way.

limit A basic concept in **calculus**, expressing the idea of tending to an ultimate but unreachable value.

As an illustration, consider the following sequence: $1, \frac{1}{2}, \frac{1}{3}, \frac{1}{4}, \ldots$ No member of this sequence will ever equal zero, but the more terms, the closer will they approximate to zero. The terms are said to approach the limit zero. This is written: $1/n \to 0$ as $n \to \infty$,

or $\lim_{n \to \infty} \left(\frac{1}{n}\right) = 0$

The notion of a limit can be illustrated geometrically.
Example
Figure L3 suggests that, although the **tangent** to the circle drawn from point P does not cut the circle, it is the limiting position of the secants

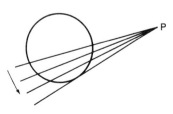

Figure L3

drawn from P that do cut the circle in two places.
See also **Achilles and the tortoise**.
LATIN *limes*: boundary

line A path having no thickness and no turning, and neither beginning nor end. It is also called 'straight line' to distinguish it from a curve, which is sometimes referred to as a 'curved line'.

In **Euclidean geometry**, only one line can be drawn through two given points.
See also **line segment, ray**.
LATIN *linea*: a linen thread

linear (a) [*lin*-i-ar] **1** Relating to **length**, as in linear measurement. **2** Relating to a straight-line graph in coordinate geometry and to expressions and functions of the first degree in algebra.
See also **linear equation, linear function**.

linear equation An equation of the first **degree**, e.g. $y = \frac{1}{2}x + 5$ (note that x has the power 1).

In coordinate geometry, the graph of a linear equation is a straight line, and every straight-line graph has a linear equation.

The general equation to a straight line is $y = mc + c$, where m is the **gradient** and c is the intercept on the y-axis.
Example
In Figure L4:

$$y = \frac{1}{2}x + 5$$
$$\text{gradient} = \frac{1}{2}$$
$$y\text{-intercept} = 5$$

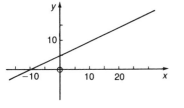
Figure L4

linear function A **function** of the first **degree**. If y is a linear function of x, then y is related to x by an equation of the form $y = ax + b$, where a and b are constants (note that x has the power 1).

Example
$\frac{1}{2}x + 3$ is a linear function of x; but the following are not linear functions of x: $1/x$, $\tan x$, $x^2 - 4$.
See also **linear equation**.

linear scale A scale along which equal distances represent equal quantities:

See also **logarithmic scale**.

line of best fit In statistics, a straight line graph that best fits a set of data.
Example
The length of a spring is measured when weights of different mass are attached to it:
    mass (g)   0   20   40   60   80   100
    length (cm) 20   26   31   37   40   46
These pairs of results are plotted on a graph and a line drawn to make the best fit, as in Figure L5.
Compare **curve of good fit**.

Figure L6

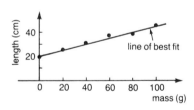

Figure L5

Figure L7

Figure L8

line of symmetry A line dividing a plane shape into two parts, each of which is a mirror image of the other (see Figure L6).
    An isosceles triangle has a line of symmetry, a rectangle has two lines of symmetry and any diameter of a circle is a line of symmetry for that circle as shown in Figures L7, 8 and 9.
    A scalene triangle has no line of symmetry.
Also called **axis of symmetry**.

Figure L9

line plot A simple way of displaying a small set of data. Also called dot plot.
Example
Figure L10 shows the line plot of the marks obtained in a test given to a class of 16 students. Each dot represents a student.

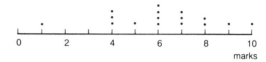

Figure L10

line segment The portion of a **line** between two points on the line. In **Euclidean geometry**, it is the shortest path between two points.

A line segment has a **finite** length, whereas a line (strictly speaking) stretches infinitely with neither beginning nor end.
Also called line interval.

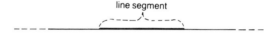

Figure L11

line segment (broken line) graph A graph in which neighbouring points are joined by **line segments**.
Example

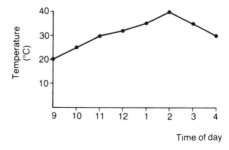

Figure L12 Graph of room temperature on a hot day

Lissajous figures [*lis*-a-zhoo *fig*-erz] Closed curves traced out by a point moving along a path resulting from two **simple harmonic motions** at right angles. They are important in the study of electronics. Named after the 19th century French physicist, Jules Lissajous.
Example
See Figure L13.

Figure L13

literal (a) Using letters (not numerals) as constants and coefficients in an algebraic expression, e.g. $ax + b$ is a literal expression, whereas $2x + 3$ is not.
LATIN *littera*: letters

litre [*lee*-ter] Abbreviation: L A non-SI unit of **capacity**.
  1 litre = 1000 millilitres (mL)
  1 litre is equivalent to 1000 cm^3.

location In statistics, a term used for **central tendency**. The main measures of location are **mean, median** and **mode**.
LATIN *locus*: place

locus [*loh*-kus or *lok*-us], (*pl.*: **loci** [*loh*-sy or *loh*-ky or *lok*-ee]) The path traced out by a point moving according to some definite rule. Most of the curves met in mathematics can be regarded as loci.
Example
A circle is the locus of a point moving in a plane and always at a constant distance (the radius) from a fixed point (the centre) in the same plane.
LATIN *locus*: a place

logarithm The logarithm of a number is the **power** to which a **base** must be raised to produce that number. The base for common logarithms is 10; the base for natural (Naperian) logarithms is **e** (≈ 2.72).
Example
Because $10^3 = 1000$, the common logarithm of 1000 is 3, written log 1000 = 3.

Because $e^3 \approx 20$, the natural logarithm of 20 is approximately 3, written ln 20 $\approx$ 3 or $\log_e 20 \approx 3$.

Logarithms may be built on any positive number (except 1) as a base. In general, if $a^x = y$, then the logarithm of y in base a is x, written $\log_a y = x$.

Logarithms have interesting properties that allow multiplication and division to be converted to addition and subtraction, and that allow powers and roots to be calculated using only multiplication and division. The rules for these conversions are:

$$\log (ab) = \log a + \log b$$
$$\log a/b = \log a - \log b$$
$$\log a^n = n(\log a)$$

Tables of logarithm values are available to help with these calculations, although their use has been largely superseded by electronic calculators. Logarithms and their properties still have important applications in advanced mathematics.

GREEK *logos*: proportion, *arithmos*: number

logarithmic function A **function** of the form $y = \log_a x$, where a is a positive constant (see Figure L14).

The **inverse** of a logarithmic function is an **exponential function**.

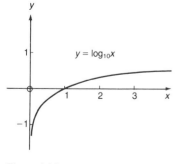

Figure L14

logarithmic scale A **scale** along which distances represent numbers in proportion to their **logarithms**.

Figure L15 shows two logarithmic scales

compared with an ordinary linear scale.

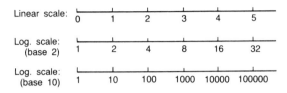

Figure L15

Logarithmic scales are used when measuring some physical quantities. For instance, the intensity of a sound increases by 1 bel for a 10-fold increase in the energy causing the sound, and a 100-fold increase in energy changes the intensity by 2 bel. (Reminder: $\log 10 = 1$, $\log 100 = 2$.)

logic [*loj*-ik] The study of the rules of reasoning. Logic is concerned with the structure of an argument rather than its content.

Mathematics has relied very much on logic, particularly as developed by Aristotle (4th century BC), **Leibniz** (17th century) and Boole (19th century). In turn, modern logic has gained through the use of mathematics and mathematical symbolism.

GREEK *logike*: art of speaking and reasoning

LOGO A computer program designed to help students learn mathematical skills. Some examples of LOGO instructions are:
FD20   move forward 20 units
BK30   move back 30 units
RT90   turn right 90 degrees
LT120   turn left 120 degrees
CIRCER20 draw a circle in a clockwise direction with radius 20 units

longitude [*lon*-ji-tyood] One of the coordinates used for locating a place on the earth's surface (the other is **latitude**). It is the angular measure between the **meridian** that

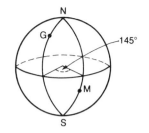

Figure L16

passes through the point and the meridian passing through Greenwich (England) (G). It is usually expressed as a number of degrees east or west (of Greenwich).
Example
The longitude of Melbourne (Australia) (M) is 145°E; this is illustrated in Figure L16.
LATIN *longitude*: length

lowest common multiple Abbreviation: LCM   The smallest **multiple** shared by two or more numbers.
Examples
4 and 6 have the following multiples:
4:   8   12   16   20   24   ...
6:   12   18   24   30   46   ...
Common multiples are 12, 24, etc.
The lowest common multiple is 12.

lowest terms   For a **fraction** to be in its lowest terms, the numerator and denominator must have no common factor (other than 1).
Example
$\frac{1}{4}$ is a fraction in its lowest terms. 10/12 can be reduced to its lowest terms by cancelling the common factor, 2, to give 5/6.

M

magic square An array of the first n^2 natural numbers arranged in a square of n rows and n columns so that every row, every column, and every diagonal adds to the same total.
Example

$n = 3$ $\qquad\qquad n = 4$

6	1	8
7	5	3
2	9	4

16	3	2	13
5	10	11	8
9	6	7	12
4	15	14	1

15 each total 34

There is only one magic square for $n = 3$ (not counting reflections and rotations), but there are 880 possibilities for $n = 4$, and 275 305 224 possibilities for $n = 5$.

magnetic north The direction pointed by the needle of a magnetic compass.
At most places on the earth's surface, magnetic north is different from true north. The angle difference between the two is known as the magnetic declination for a given place. It is important in navigation.

magnitude The size of a quantity, measured in units of that quantity, e.g. the magnitude of each angle of a regular hexagon is 120 degrees or $\frac{2\pi}{3}$ radians or $1\frac{1}{3}$ right angles.
LATIN *magnitudo*: greatness

major Greater, as in Figure M1.
Opposite: **minor**.
LATIN *major*: greater

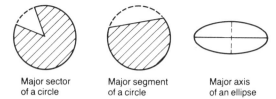

Major sector of a circle   Major segment of a circle   Major axis of an ellipse

Figure M1

mantissa The fractional part of a **common logarithm**. It is always positive.
Examples

Number	Logarithm	Mantissa
200	2.3010	.3010
20	1.3010	.3010
2	0.3010	.3010
0.2	$\bar{1}.3010$	.3010

See also **characteristic**.
LATIN *mantissa*: an addition

mapping The relating of one set of points to another set of points, or the relating of object to image.

In geometry, **dilatations, rotations, reflections** and **translations** are mappings of one figure or region into another.

In algebra, if y is expressed as a **function** of x, the calculation of values of y corresponding to values of x is a mapping of x into y.

A special case of mapping is the representation on plane paper of the curved surface of the earth. There are several techniques for achieving this; the most common is known as a **Mercator projection**.

mass A basic concept related to the amount of matter in an object. The greater the mass of an object, the harder it is to change its motion or to start it moving.

In everyday use, masses are compared by comparing weights; but, whereas the weight of

an object depends on how close it is to the earth (or other large object in the universe), the mass does not change with position.
The **SI** standard unit of mass is the kilogram (kg), which is the mass of a special lump of metal kept in France.

mathematics *See* Introduction.
GREEK *mathema*: learning

matrix [*may*-triks], (*pl.*: **matrices** [*may*-trseez])   A square or rectangular array of numbers set out in rows and columns.
Example
$\begin{vmatrix} a & b \\ c & d \end{vmatrix}$ is a square 2 × 2 matrix.
In matrix algebra, there are rules for adding and multiplying matrices.
Examples
$$\begin{vmatrix} a & b \\ c & d \end{vmatrix} + \begin{vmatrix} e & f \\ g & h \end{vmatrix} = \begin{vmatrix} (a+e) & (b+f) \\ (c+g) & (d+h) \end{vmatrix}$$

$$\begin{vmatrix} a & b \\ c & d \end{vmatrix} \times \begin{vmatrix} e & f \\ g & h \end{vmatrix} = \begin{vmatrix} (ae+bg) & (af+bh) \\ (ce+dg) & (cf+dh) \end{vmatrix}$$

Matrices are important in the study of **vectors** and in the solution of **simultaneous equations**.
LATIN *matrix*: womb

maximum, (*pl.*: **maxima**)   The greatest value.
Example
The greatest value of $3 - x^2$ is 3, because x^2 cannot have any value less than zero. So $3 - x^2$ has a maximum value of 3 (see Figure M2).

Some functions have a maximum that is the greatest value only for a certain region (see Figure M3). Some functions have several maxima (see Figure M4), and some functions have no maxima (see Figure M5). In all the graphs, the greatest values (the maxima) are labelled M on the *y*-axis.
LATIN *maximus*: greatest

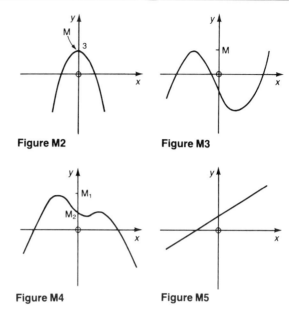

Figure M2

Figure M3

Figure M4

Figure M5

mean *See* **arithmetic mean, geometric mean, harmonic mean.**
See also **median, mode.**
LATIN *medius*: middle

measure The precise indication, in units, of the size of something. For instance, for a geometric figure, there are measures of length, area, volume, angle — expressed in metres, square metres, cubic metres, degrees, respectively, and other units.

Linear measure refers to measurement of length; square measure refers to measurement of area; cubic measure refers to measurement of volume.
LATIN *mensura*: a measurement

measurement 1 The act of finding a measure. 2 Another name for the measure itself.
See also **approximation, error.**

median [*mee*-di-an]
1 (arithmetic)—The middle term of a sequence of numbers arranged in ascending order, or the average of the two middle terms if there is an even number of terms.
Example
The median of each of the following sequences is 10:
  4, 6, 10, 12, 19
  1, 3, 8, 12, 14, 20.
2 (statistics)—The middle term of a **frequency distribution**, meaning that the median divides the total distribution into two equal parts.
Example
The median of the following distribution is 1.5, since 50% of all cases fall below this value and 50% above:

x	0	1	2	3	4	5	6
frequency	10	40	29	9	8	4	0

Median is an example of a measure of **central tendency**. Others are **mean** and **mode**.
3 (geometry)—Line joining a **vertex** of a triangle to the midpoint of the opposite side. A median divides the triangle into two equal areas. Every triangle has three medians. They are **concurrent** at a point called the **centroid** (see Figure M6).
LATIN *medianus*: in the middle

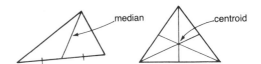

Figure M6

mega- A prefix meaning one million times. Symbol M.
Example
  1 megalitre = 1 000 000 litres
  1 ML = 10^6 L
GREEK *megas*: great

mensuration Part of mathematics that deals with the measurement of length, area and volume.
LATIN *mensura*: measure

Mercator projection [mer-*kay*-tor pro-*jek*-shun] One way of **mapping** the spherical surface of a globe representing the earth onto a plane surface. It is the **projection** of the globe on to a **cylinder**, which is then opened out flat (see Figure M7).

On the map, parallels of **latitude** and meridians of **longitude** are straight lines and cross each other at right angles just as on the globe. Shapes are shown correctly. However, distances and areas towards the poles appear stretched compared with distances and areas near the equator.

Named after a 16th century Flemish mapmaker.

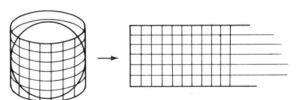

Figure M7

meridian [mer-*id*-i-an] In geography, a semicircle joining the north and south poles and passing through a given point on the earth's surface. Meridians on a map are labelled with the number of degrees east or west of the meridian passing through Greenwich (England), e.g. the meridian through Auckland (NZ) is 175°E.

All meridians are at right angles to parallels of **latitude**, and they meet at the north and south poles.
See also **longitude**.
LATIN *meridies*: midday

metre [*mee*-ter] The standard or base unit of length in the international system of units (**SI**). The metre was first defined in 1795 as one ten-millionth of the distance between the equator and the north pole measured along the **meridian**, passing through Paris. In 1889 it was redefined as the distance between two marks on a special bar kept near Paris. It is now defined in terms of the wavelength of radiation from a krypton atom.
FRENCH *mètre*, from Greek *metron*: measure

metric system Any system of measurement based on multiples and submultiples of ten and including the metre as a unit of length.
The **Système international d'unités (SI)** is the system usually intended by this term.

micro- A prefix meaning one-millionth. Symbol μ (Greek letter pronounced mu).
Example
$$1 \text{ microgram} = 0.000\,001 \text{ gram}$$
$$1 \text{ μg} = 10^{-6} \text{ g}$$
GREEK *mikros*: small

midpoint The point on a **line segment** that divides it into two equal parts.

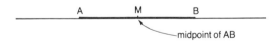

Figure M8

milli- A prefix meaning one-thousandth. Symbol m.
Example
$$1 \text{ millimetre} = 0.001 \text{ metre}$$
$$1 \text{ mm} = 10^{-3} \text{ m}$$
LATIN *mille*: a thousand

minimum (*pl.* **minima**) The least value.

mixed number

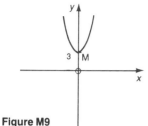

Figure M9

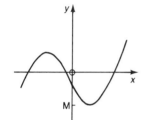

Figure M10

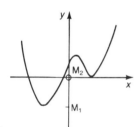

Figure M11

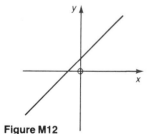

Figure M12

Example
The least value of $x^2 + 3$ is 3, because x^2 cannot have any value less than zero; so $x^2 + 3$ has a minimum value of 3 (see Figure M9).

Some functions have a minimum that is the least value only for a certain region (see Figure M10), whereas some functions have several minima (see Figure M11), and some functions have no minimum (see Figure M12). In all the graphs, the minima are labelled M on the y-axis.

LATIN *minimus*: least

minor (a) Smaller, as in the diagrams shown in Figure M13.
Opposite: major.
LATIN *minor*: less

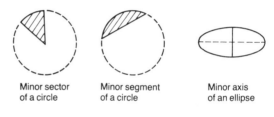

Minor sector of a circle

Minor segment of a circle

Minor axis of an ellipse

Figure M13

minus sign The symbol '−'. It has two main uses: **1** To indicate subtraction, e.g. $10 − 7 = 3$, read 'ten minus seven equals three'. **2** To mark a number less than zero, e.g. -2, read 'minus two' or 'negative two'.
See also **directed number**.

minute 1 An angle measure equal to one-sixtieth of a degree. **2** A time measure equal to 60 seconds or one-sixtieth of an hour. Symbol $'$, e.g. 15$'$ means 15 minutes.

mixed number A number expressed as the sum of a whole number and a proper fraction, e.g. $2 + \frac{1}{2}$, shortened to $2\frac{1}{2}$. Any improper fraction can be converted to a mixed number, e.g. $7/4 = 1\frac{3}{4}$.

Möbius strip [*mer*-bi-us] A special twisted surface with only one side and one edge.
To construct an example, take a long narrow rectangular strip of paper, give it a half twist and join the ends together. Figure M14 shows a möbius strip contrasted with a cylinder made from a similar paper strip.
Named after August Möbius, a 19th century German mathematician.

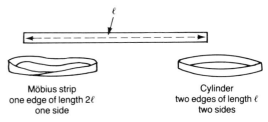

Möbius strip	Cylinder
one edge of length 2ℓ	two edges of length ℓ
one side	two sides

Figure M14

mode Of a set of observed values in statistics, the value that occurs most often.
Example
The ages (in years) of members of a club are recorded as 12,12,13,12,14,15,13,13,13,11,14. The mode is 13, because it is the age that appears more often than any other.
The mode is an example of a measure of **central tendency**. Others are **mean** and **median**.

modelling In applied mathematics, the technique of producing a mathematical description to help solve a practical problem.
In the simplest cases, the mathematical description or model may be a graph or algebraic function expressing the relation between two variables, e.g. the relation between pulse rate and time elapsed since rapid exercise. In more complex cases, there may be many variables whose relationships are to be described

by complicated mathematical functions.
The problems that can be attacked by mathematical modelling come from almost any field. A few examples are: electrical networks, moon landing, chess, music, sports, population, taxation, medical science, traffic flow. Computers have extended the speed and range of complexity of modelling.

modular arithmetic A system of arithmetic in which two numbers are considered equivalent (**congruent**) if they give the same remainder when divided by a fixed number called the **modulus**.
Example
If the modulus is 12, then the numbers 14 and 26 are congruent because the remainder after dividing by 12 is 2 in each case. In other words, 14 and 26 differ by an integral number of times the modulus: $14 = 1 \times 12 + 2$, $26 = 2 \times 12 + 2$.

When the modulus is 12, the rules for adding and subtracting are like the rules for adding and subtracting hours on a 12-hour clock.

The modulus may be any number e.g. if the modulus is 5, then 16 and 21 are congruent. This is written $16 \equiv 21 \pmod 5$.

Numbers that, in decimal notation, have the same units digit are equivalent, modulus 10. A similar rule can be stated for numbers in **binary**, **octal** or **hexadecimal** notation, e.g. $10101 \equiv 1011 \pmod{10}$. This statement, as it stands, holds true whether it is read in binary (10 = two), octal (10 = eight), hexadecimal (10 = sixteen) or decimal (10 = ten).

modulus
1 The **absolute** value of a number: its value without regard for sign, e.g. $+7$ and -7 have the same modulus, 7. The modulus or absolute value of n is written $|n|$ and is read 'mod n'.
2 The base of a system of **modular arithmetic**, e.g. 10 and 14 are equivalent in modulus

arithmetic to modulus 4 (the remainder is 2 after each is divided by 4), but not if the modulus is 5 (then the remainders are 0 and 4).
LATIN *modulus*: a small measure

mu μ, the twelfth letter of the Greek alphabet, corresponding to m. μ is used in statistics for the **mean** of a **population**.

multiple
1 Multiples of a whole number can be found by multiplying it by any whole number other than 1.
Example
Some multiples of 5 are 10, 15, 20, 25 (formed by multiplying by 2, 3, 4, 5). The multiple of a number always has that number as a **factor**.
2 Similarly in algebra a **polynomial** has multiples.
Example
A multiple of $a + b$ is $(a + b)(a - b) = a^2 - b^2$. $a + b$ is a factor of $a^2 - b^2$.
LATIN *multiplus*: many

multiplicand *See* multiplication.

multiplication One of the **binary operations** of mathematics. In its simplest form, multiplication is the process of adding a whole number to itself a certain number of times, e.g. the multiplication of 5 by 3 is $5 + 5 + 5$, producing the **product** 15. In this case, 5 is called the multiplicand and 3 is the multiplier. The process is written $5 \times 3 = 15$.

Multiplication is extended to include non-integers, using the **associative** and **commutative** properties. It is then **closed** for all **real numbers**.

In algebra, if p is multiplied by q, the product is shown as $p \times q$ or $p.q$ or pq.

The **inverse** of multiplication is **division**.

Rules for the multiplication of two signed numbers:
- If two numbers have the same sign, their product is positive, e.g. $+4 \times +5 = +20$; $-4 \times -5 = +20$.
- If two numbers have different signs, their product is negative, e.g. $+4 \times -5 = -20$; $-4 \times +5 = -20$.

LATIN *multiplicare*: to make many folds

multiplier See **multiplication**.

multiply (v) To combine one number with another by **multiplication**, giving a **product**.

mutually exclusive In statistics, two or more **events** or outcomes that cannot occur together.
Example
If a coin is tossed once, the result is one of two possibilities that are mutually exclusive. The outcome is either a head or a tail, but not both. The **probability** of one of these occurring is $\frac{1}{2}$; the probability that the result is a head *or* a tail is found by adding the separate probabilities—$\frac{1}{2} + \frac{1}{2} = 1$ (certainty).

Events that are not mutually exclusive may be **independent**. If a coin is tossed twice, the outcome of the second toss is independent of the outcome of the first: the two events are not mutually exclusive. The probability of a head *and* a tail in this case is found by multiplying the separate probabilities—$\frac{1}{2} \times \frac{1}{2} = \frac{1}{4}$.

N

nano- [*nan*-oh]   A prefix meaning one-thousand-millionth. Symbol n.
Example
1 nanosecond = 0.000 000 001 second
1 ns = 10^{-9}s
GREEK   *nanos*: a dwarf

Napier, John (1550–1617)   A Scottish baron who spent much of his spare time working at mathematics. He is remembered for the invention of **logarithms**. He also invented a set of rods, known as Napier's bones, to help in multiplication and division. These were the forerunner of the **slide rule**.

natural logarithm   A **logarithm** using base **e**. Also called Naperian logarithm.
   The natural logarithm of x is written $\log_e x$ or $\ln x$.

natural number   One of the counting numbers, one, two, three, four, etc. (zero is sometimes included). A natural number is used as a **cardinal** number when it describes how many things there are in a set (ten runners in a race), and as an **ordinal** number when it marks the position of something in a sequence (the runner in place number ten).
   For other kinds of number, see **imaginary number, number, rational, real number**.

nautical mile   An international unit of distance used in sea and air navigation. It is defined as 1.852 kilometres. Originally, it was defined as 1 degree of **latitude**.
   The knot is a unit of speed equal to 1 nautical

mile per hour.
GREEK *nautikos*: of ships or sailors

negative (a)
1 Of a **directed number**, one that is less than zero. On the **number line**, negative numbers are usually to the left of zero, while **positive** numbers are to the right. A negative number is indicated by a **minus sign**, e.g. negative 3 (or minus 3) is written ⁻3 or −3.
2 Of the sense of a direction, the opposite to positive.

Positive and negative angles are measured in anticlockwise and clockwise rotations respectively.

A negative acceleration is an acceleration in the opposite sense from the motion of the object; that is, a slowing down.

For negative correlation, *see* **correlation**.
LATIN *negare*: to say no

net A flat diagram consisting of the plane faces of a **polyhedron**, arranged so that the diagram may be folded up to form the solid.
Example
The diagrams in Figure N1 are nets of a cube and of a square pyramid.
See also **develop**.

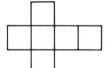

Figure N1

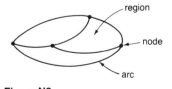

Figure N2

network In **topology**, a diagram consisting of arcs (branches), nodes (junctions) and regions, as in Figure N2.

Networks have many applications, such as in electrical circuitry and chemistry.
See also **Euler's formula**.

Newton, Isaac (1643–1727)   An English scientist and mathematician, regarded by many as the greatest scientific genius of all time. Newton's most important contribution to mathematics was the invention of the **calculus**, which he called his method of 'fluxions'. (The calculus was also claimed to have been invented about the same time by **Leibniz**.)

nine-point circle   The circle, associated with a triangle, that passes through the following nine points:
- the mid-points (a) of the sides
- the points (b) where the altitudes meet the sides
- the points (c) midway between the vertices and the point of intersection of the altitudes.

The radius of this circle equals half the radius of the **circumcircle**.

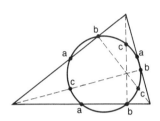

Figure N3

nominal data   Information that can be counted in **discrete** categories without being otherwise measured.
Example
A school list showing various subjects and the students enrolled for them. The subjects may be labelled by numbers, but these numbers do not suggest any order or size and they cannot enter into any mathematical calculations. They are merely labels.
Compare   **ordinal data**.
LATIN   *nominare*: to call by name

nonagon [*non*-a-gon]   A nine-sided **polygon**. Each interior angle of a regular nonagon measures 140°.
Also called enneagon [*en*-i-a-gon].
LATIN   *nonus*: ninth,   GREEK   *gonia*: angle

non-euclidean geometry   Geometry not based on the **axioms** of **Euclid**. Several such geometries have been developed in the past two centuries. The geometries of Riemann (1826–66) and Lobachevski (1793–1856) do not accept Euclid's axiom about parallels. This axiom is that, given a line and a point not on it, then there is one line that can be drawn through the point parallel to the given line. Riemann's geometry assumes there is no such line, and Lobachevski's geometry is based on the assumption that they are many.

Geometries like these lead to different understandings about **space**. Einstein's relativity theory used Riemann's geometry.

normal (a) At right angles; **perpendicular**. The word is also used as a noun to name a line that is at right angles to another line or to a plane.

The normal to a curve at a given point is the line at right angles to the tangent of the curve at that point (see Figure N4).

The normal to a curved surface at a given point is the line at right angles to the tangent plane at that point.

LATIN *norma*: a carpenter's square

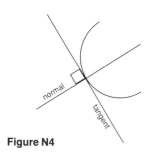

Figure N4

normal curve A curve of special importance in statistics and probability theory. The curve is symmetrical and bell-shaped and represents the probability distribution (f) plotted vertically for a random variable (x) plotted horizontally.

In Figure N5, the horizontal axis is marked in **standard deviation** units and the area of the

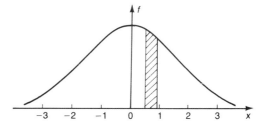

Figure N5

normal distribution

section that has been shaded as an example is a measure of the probability that x has the values lying within the limits of that section. Tables of probabilities are published in statistics textbooks.
See also **normal distribution**.

normal distribution A **frequency distribution** that can be represented by the **normal curve**. **Histograms** representing many actual frequency distributions of scientific and natural events approximate to a normal distribution, and this fact is used for making predictions.
Example
The distribution of the following variables approximate to a normal distribution: heights of people, times for athletes to run 5000 metres, life expectancy of light bulbs, and so on.

notation Any system of signs and symbols invented for a special purpose and different from ordinary writing. The decimal way of writing numbers is one of the most important notations introduced into mathematics. Nearly all parts of mathematics use notations for making statements and indicating operations, just as in music a notation is used to represent musical notes and rhythms on paper.
LATIN   *notatio*: a marking

null set The **empty set**. The empty set has no members. There is therefore only one possible null set. It is represented in set theory as { } or ∅.

number The following account of the various kinds of number shows some of the logic underlying mathematics. It is not intended as a history of mathematics.

Numbers at their simplest are the **natural numbers** (or whole numbers) used for counting (1,2,3, etc.) The set of whole numbers is **closed** under addition and multiplication, meaning that any whole numbers added or multiplied together produce a whole number.

The number system can then be extended to include all **integers**: positive integers (1,2,3, etc.), **zero** (0), and negative integers ($-1, -2, -3$, etc.). The set of integers is closed under subtraction as well as addition and multiplication.

The system can next be extended to include **fractions** (e.g. $\frac{1}{2}$). It is now the set of **rational numbers**, and is closed under division as well as addition, multiplication and subtraction.

When **irrational** numbers (e.g. $\sqrt{2}$) are included, we have the set of **real numbers**, all of which can be represented on the **number line**.

A further step is taken in higher mathematics with the inclusion of numbers that involve the square roots of negative numbers. When these are included we have the set of **complex numbers**.

Numbers are represented by symbols known as **numerals**, though the word 'number' often serves that purpose too.

number line A line on which all the **real numbers** can be represented by points at distances from a fixed point representing **zero**.
Example

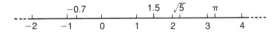

Figure N6

numeral   A symbol representing a particular **number**.
Example
The number eight is represented by several different numerals: 8 (decimal system), 1000 (binary system), VIII (Roman), 八 (Chinese).
See also   **Arabic numerals, pronumeral**.

numerator Of a **fraction** written in the form $\frac{a}{b}$, the number above the line.
Examples
The numerator of $\frac{3}{4}$ is 3; the numerator of $(x^2 - 1)/(x + 1)$ is $x^2 - 1$.
LATIN *numerus*: a number

numerical (a) **1** In algebra, referring to the presence of **numerals** rather than **pronumerals**, e.g. $2x + 3$ is a numerical expression, whereas $ax + b$ is a **literal** expression. **2** Sometimes used as another word for **absolute**, e.g. the numerical value of both $+5$ and -5 is 5.

O

object — see **image**.

oblate (a) [*oh*-blayt]   Flattened. An oblate **spheroid** is formed by rotating an **ellipse** around its minor axis.
Example
The earth has the shape of an oblate spheroid: places at the equator are further from the centre of the earth than a place at either the north or the south pole.

oblique (a) [oh-*bleek*]   Neither **perpendicular** to nor **parallel** to a given line or plane.
Example
In Figure O1, AB and PQ are oblique to each other, and XY is oblique to the plane shown.

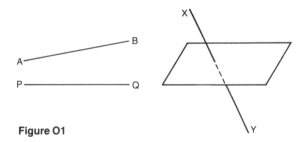

Figure O1

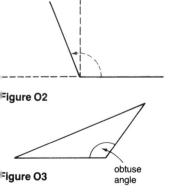

Figure O2

Figure O3

obtuse angle

obtuse (a)
1 An obtuse angle is greater than a right angle (90°) but less than two right angles (180°) (see Figure O2). **2** An obtuse-angled triangle has one of its angles greater than a right angle (see Figure O3).
LATIN   *obtuse*: blunt

occurrence   Another name for an **event**.

octagon An eight-sided **polygon**. Each interior angle of a regular octagon measures 135°.
GREEK *okto*: eight, *gonia*: angle

Figure O4

octahedron A solid figure with eight plane faces. A regular octahedron is one of the five possible shapes for a **regular** polyhedron. Its faces are equilateral triangles.
GREEK *okto*: eight, *hedra*: base

octal (a) Of a number system, based on eight.
Example
In the octal system, the symbol 23 represents two eights + three units. (In the decimal system, it represents two tens + three units.)
So $23_{eight} = 19_{ten}$.
The digits 8 and 9 are not used in the octal system. Eight is written 10, eight squared is written 100, eight cubed is 1000 and so on.
Compare **binary notation, hexadecimal**.
GREEK *okto*: eight

odd (a) 1 Odd number. A whole number that is not **even**. When divided by 2, the remainder is 1, e.g. 7, 15, −3 are odd numbers. If *n* is any integer (odd or even), then $2n + 1$ must always be odd.
2 Odd function. If f(*x*) is an odd **function** of *x*, then replacing *x* by −*x* changes the sign of f(*x*) but not its **absolute** value.
Example
x^3 is an odd function of *x*, because $(-x)^3 = -x^3$.
The graph of an odd function has **rotational symmetry** about the origin (see Figure O5).
See also **even**.

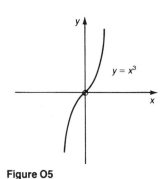

Figure O5

odds The **probability** that an event will occur compared with the probability of its not occurring. This comparison is usually expressed as a **ratio** in favour or against.
Example
If two coins are tossed, the possible outcomes

are HH, HT, TH, TT. Therefore the probability of two heads appearing in one toss is 1/4 and the probability of two heads not appearing is 3/4. The odds are expressed as 1 to 3 in favour or 3 to 1 against.

ogive [*oh*-jyv]   In statistics, the graph obtained from a **cumulative frequency** distribution. It usually has the shape of an elongated S.
Example
Figure O6 shows the curve obtained from the example included under the heading, **cumulative frequency**.

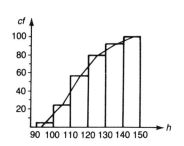
Figure O6

FRENCH   An architectural term

one-dimensional   Referring to the geometrical condition that requires only one coordinate (or **dimension**) to specify the position of a point.

If a variable point is restricted to a given line, then its position can be stated using only one number, namely its displacement from a reference point on the line. A line is thus described as a one-dimensional **space**.
Compare   **three-dimensional, two-dimensional**.

one-to-one correspondence   A relationship between two **sets** such that each element of each set can be linked with a single element of the other. Sets having this relationship are **equivalent**.
Example
The set of values of x and the set of values of $2x - 1$ can be put into one-to-one correspondence, because for every value of x there is one value of $2x - 1$, and for every value of $2x - 1$ there is one value of x. This is neatly illustrated by the graph of $2x - 1$ against x in Figure O7.

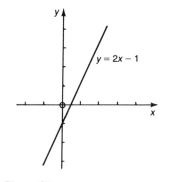
Figure O7

operation   A mathematical process combining numbers or sets.

order

The fundamental operations of arithmetic are addition and multiplication with their **inverses**, subtraction and division. Other operations include extracting roots, union of sets, intersection of sets. Each operation has its own special symbol, e.g. + (addition), − (subtraction), × (multiplication), ÷ (division), √ (root extraction), ∪ (union), ∩ (intersection).
LATIN *operari*: work

order 1 Ascending or descending order. A sequence of numbers written from smallest to largest or largest to smallest, e.g.
7,9,11,13... ascending order
13,11,9,7,... descending order
2 Order of a square **matrix**. The number of rows in the matrix. **3** Order of **rotational symmetry**. The number of times a shape coincides with its original position as it turns through one complete rotation, e.g. an equilateral triangle has rotational symmetry of order three. **4** Order of magnitude. The approximate size of something, usually expressed in powers of ten, e.g. 1320 and 1500 may, for some purposes, be thought to have the same order of magnitude, but 14370 has an order of magnitude ten times greater.

ordered pair A pair of numbers for which their order is important. The **coordinates** of a point in the *x*–*y* plane form an ordered pair, e.g. (3,5) corresponds to point P, but (5,3) corresponds to point Q in Figure O8.

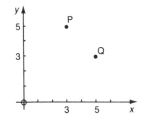

Figure O8

ordinal data Information that can be ranked in order, but without further measurement.
  The ranking of teams in a sporting contest provides an example of ordinal data; for, although one team can be judged better than another, it is not usually possible to say how many times better.
Compare **nominal data**.

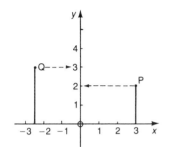

Figure O9

ordinal number A number indicating the position of a thing in a set of things arranged in sequence.
Example
The runner who comes third in a race has the ordinal number 3. On the other hand, to say that there were 12 people in the race is to quote the **cardinal** number of the set of runners. Ordinal numbers are often written 1st, 2nd, 3rd, etc.

ordinate On a **Cartesian** graph, the distance of a point from the *x*-axis; the *y*-coordinate.
Example
In Figure O9, the ordinate of P is 2, and the ordinate of Q is 3.
See also **abscissa**.

origin The point from which a measurement is taken; especially the intersection of two or more axes in a system of coordinates.
LATIN *origo*: beginning

orthocentre The point where the three **altitudes** of a triangle intersect. It is a property of any triangle that its three altitudes are **concurrent**. For some, the point of intersection is inside the triangle and for some it is outside the triangle. Figure O10 illustrates this.
GREEK *orthos*: straight

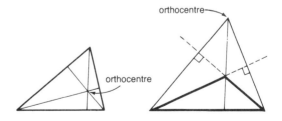

Figure O10

orthogonal (a) [aw-*thog*-on-l] Describing lines or planes that are at right angles to each

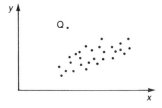

Figure O11

other; **normal; perpendicular**.

GREEK *orthos*: straight, *gonia*: angle

outlier [*owt*-ly-er] In statistics, a point in a sample separated from the main body of the sample, e.g. in Figure O11, point Q in the **scattergram** is an outlier.

P

palindromic (a) [pal-in-*droh*-mik] Read the same backwards as forwards. The following numerals are palindromic: 121, 4774, 50105.
GREEK *palin*: back, *drom*: run

pantograph An instrument for drawing one plane figure similar to another on smaller or larger scale. The scale can be fixed at various values.
GREEK *pan*: all, *graphe*: drawing

parabola [pa-*rab*-o-la] One of the **conic sections**. It is the curve formed when a plane cuts a right circular cone parallel to the sloping surface of the cone (see Figure P1).

It is also the **locus** of a point moving so that its distance from a fixed point (the focus) equals its distance from a fixed line (the directrix) (see Figure P2).

If a parabola is drawn on a Cartesian graph so that it is symmetrical about the *y*-axis with its vertex at the origin (see Figure P3), its equation

Figure P1

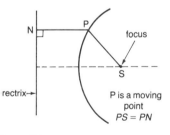

Figure P2

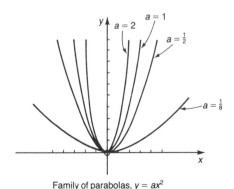

Family of parabolas, $y = ax^2$

Figure P3

is $y = ax^2$ where a is a constant.

GREEK *parabole*: comparison

parabolic (a) [pa-ra-*bol*-ik]   Describing the shape of a **parabola**, e.g. a fast jet of water from a hose follows approximately a parabolic path if there is no wind.

paraboloid [pa-*rab*-o-loyd]   In its simplest form, the surface or solid formed when a parabola is rotated about its axis.

Many reflectors (such as for radar, searchlights, satellite dishes) have the shape of a paraboloid and use the following property: parallel rays striking the inside of a paraboloid are focused to a point, and, conversely, rays from that point emerge after reflection as a parallel beam, as illustrated in Figure P4.

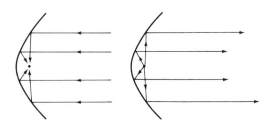

Figure P4

paradox [*pa*-ra-doks]   A self-contradictory statement. It may at first sight appear logical, while leading to an absurd conclusion. For an example relevant to mathematics, see **Achilles and the tortoise**.

GREEK *paradoxus*: contrary to expectation

parallel   Of two lines—Lying in the same plane and never meeting however far produced. Of two planes—Everywhere the same distance apart.

One of **Euclid's axioms** was that, through a

given point not on a given line, there is exactly one other line that can be drawn parallel to the first. In some modern mathematics, this axiom has been abandoned, making way for **non-euclidean geometry**.
See also **skew**.
GREEK *para-*: beside, *allelo*: one another

parallelepiped [*pa*-ra-lel-e-*py*-ped] A six-faced solid with every face a **parallelogram**. A **cube** is a special case of a parallelepiped.
See also **prism**.
GREEK parallel + *epipedon*: flat surface

parallelogram [pa-ra-*lel*-o-gram] A **quadrilateral** with opposite sides parallel (see Figure P5).
Parallelograms are plane figures. Their opposite sides are equal as well as parallel, their opposite angles are equal and their diagonals bisect each other. Some special parallelograms are: **rectangle** (all angles equal), **rhombus** (all sides equal and diagonals at right angles), **square** (all sides equal and all angles equal).
GREEK parallel + *gramma*: something written

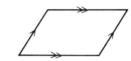
Figure P5

parameter [pa-*ram*-e-ter]
1 In algebra, a constant that can have different values in an expression without changing the form of the expression.
Example
Figure P6 shows two graphs corresponding to the equation $y = mx$. In each case, the graph is a straight line passing through the origin; the different slopes correspond to different values of the parameter, m.
2 In statistics, a constant (such as **mean**) that is characteristic of a whole **population**, as distinct from a **statistic**, which is characteristic of only a **sample** of the population.
Greek letters are often used for population parameters, and roman letters for sample

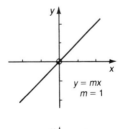
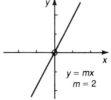
Figure P6

statistics, e.g. µ for population mean, m for sample mean, σ and s for the corresponding standard deviations.

GREEK *para*: beside, *metron*: measure

parenthesis [pa-*ren*-the-siz] (*pl.* **parentheses** [pa-*ren*-the-seez] Used in the plural, round **brackets**: ( ).

GREEK *para-*: beside, *thesis*: a placing

Pascal [pas-*kahl*], **Blaise** (1623–62) A French mathematician, physicist and theologian. In mathematics, he is famous for having laid the foundation for the theory of probability. At the age of twelve, he discovered for himself several basic theorems in geometry. At sixteen, he wrote an important essay on conics, and at nineteen invented an adding machine.

Pascal's triangle A triangular array of whole numbers formed in such a way that each number is the sum of the two numbers immediately above it (see Figure P7).

```
              1
            1   1
          1   2   1
        1   3   3   1
      1   4   6   4   1
    1   5  10  10   5   1
              etc.
```

Figure P7

The triangle was published by **Pascal** in 1665, but it was well known in Europe before that. It appeared in Chinese books as a means of calculating the **binomial coefficients** as long ago as the year 1100.

The rows of numbers in the triangle form the coefficients in the binomial expansion, $(a + b)^n$, e.g. $(a + b)^3 = 1a^3 + 3a^2b + 3ab^2 + 1b^3$

Peano [pay-*ah*-noh], Guiseppi (1858–1932) An Italian mathematician who, using symbolic logic, attempted to establish all mathematics as a deductive system derived from basic principles.

pentagon A five-sided **polygon**. Pentagons are plane figures with straight sides. A regular pentagon is one with equal sides and equal angles. Each angle of a regular pentagon measures 108°.
GREEK *pente*: five, *gonia*: angle

Irregular pentagons   Regular pentagon

Figure P8

pentagram A star-shaped figure formed by drawing the diagonals of a regular **pentagon**. It is also formed by extending the sides of a regular pentagon until they meet.
Also called pentacle or pentangle.
GREEK *pente*: five, *gramma*: something written

Figure P9

Figure P10

pentahedron [pen-ta-*hee*-dron] A solid figure with five plane faces. A square **pyramid** is an example of a pentahedron (Figure P10). There is no regular pentahedron.
GREEK *pente*: five, *hedra*: base

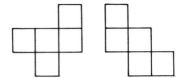

Figure P11

pentomino [pen-*tom*-in-oh]   A plane figure made from five equal squares so that some sides are shared.
Examples   See Figure P11.

per cent   An abbreviation for Latin words meaning 'by the hundred' or 'out of a hundred', used in expressions such as 'the rate of interest is 18 per cent'. Symbol %. 18% stands for 18/100.
See also   **percentage**.

percentage   A rate or a proportion expressed as part of one hundred.
Example
The percentage of the population who are male is 50. This means that 50 of every 100 persons chosen at random, or one-half of the population, are males. This proportion is written 50% (read 'fifty percent').
   Any percentage can be expressed as a fraction (either common or decimal), and any fraction can be expressed as a percentage.
Examples
a   25% = 1/4 = 0.25
b   80% = 4/5 = 0.8
c   150% = 3/2 = 1.5
LATIN   *per centum*: by the hundred

percentile [per-*sen*-tyl]   One of the 99 values of a variable dividing a distribution into 100 integral percentage parts.
Example
If a school population is ranked in order of height and it is discovered that 80% of the population have heights no greater than 145 centimetres and the other 20% are taller than this, then 145 centimetres is at the 80th percentile of the height distribution.
   The height that divides the population into two equal parts is at the 50% percentile and is called the **median** height. The upper and lower **quartiles** occur at the 25th and 75th percentiles,

and the range of heights between these two quartiles is the **interquartile range**.

perfect (a)
1   A perfect number is a **whole number** that is the sum of all its own different factors, except itself.
Examples
a $6 = 1 + 2 + 3$
b $28 = 1 + 2 + 4 + 7 + 14$
c $496 = 1 + 2 + 4 + 8 + 16 + 31 + 62 + 124 + 248$
2   A perfect square is a number that can be expressed as the product of two equal factors.
Examples
$9 = 3 \times 3; 289 = 17 \times 17; x^2 - 2x + 1 = (x - 1)(x - 1)$

perigon   The angle equal to a complete rotation. Its measure is 360 degrees or four right angles or 2π radians.
GREEK   *peri-*: around, *gonia*: angle

perimeter   The boundary of a plane figure, or the length of this boundary, e.g. the perimeter of a rectangle is the sum of its four sides; the perimeter of a circle is its circumference.
GREEK   *peri-*: around, *metron*: measure

period   The time taken for a repeating action to complete one cycle.
Example
The period of the earth's movement around the sun is 1 year. If a heart beats 60 times a minute, the period of the heartbeat is 1 second.
   The word is also applied to an **interval** of values of the independent variable of a **periodic function**, even when time is not involved.
GREEK   *peri-*: around, *hodos*: a way or a path

periodic function   A **function**, f(x), whose values are repeated at equal intervals of the

independent variable, x, as x continues to increase.
Example
Figure P12 illustrates a periodic function with a period of 2 units. Important examples of a periodic function are the sine, cosine and tangent functions.

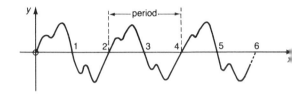

Figure P12

permutation An arrangement of a set of things in a particular order.
Example
For the letters A, B, C, one permutation is ABC and another is BAC. There are six possible permutations of these three letters: ABC, ACB, BCA, BAC, CBA, CAB.

For n things, the number of permutations is n *factorial* ($n!$). This means, for example, that the number of ways of arranging five people in a line is $5 \times 4 \times 3 \times 2 \times 1 = 120$.

If r things are to be selected from a total set of n things, then the number of permutations can be calculated by the formula, $n!/(n-r)!$ This can show that there are 60 arrangements of three people chosen from a set of 5.

LATIN *permutare*: to change

perpendicular (a) [per-pen-*dik*-yoo-lar] (sometimes used as a noun) At right angles.
Example
In Figure P13, the tangent and radius are perpendicular to each other. In Figure P14, the axis of the right circular cone is perpendicular to

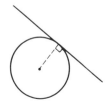

Figure P13

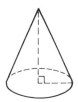

Figure P14

the base. The axis is the perpendicular drawn to the base.

See also **normal**.

LATIN *perpendiculum*: a plumb-line

perpendicular bisector A line drawn **perpendicular** to a line segment and dividing it into two equal parts. Any point on the perpendicular bisector of a line segment is **equidistant** from the two end-points of the line segment.

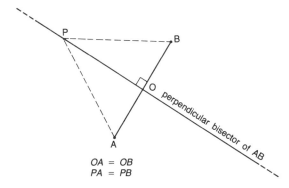

Figure P15

perpendicular distance The distance of a point from a line, measured along a line perpendicular to the given line. It is shorter than any other distance that can be measured from the point to the line.

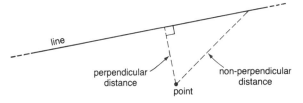

Figure P16

perspective [per-*spek*-tiv] A method of drawing objects on a flat surface so that they

appear to be in three **dimensions**.
Example
In Figure P17, two points are marked on a line representing the horizon, and lines that are to appear **horizontal** (like the edges of the box) are made to converge towards these points. **Vertical** lines are kept vertical.

The method of perspective was developed by Italian artists in the 15th century.

LATIN   *per-*: through; *specere*: to look

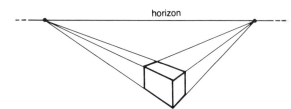

Figure P17

peta- [*pee*-ta]   A prefix meaning one thousand million million times. Symbol P.
Example
1 petametre = 1 000 000 000 000 000 metres
1 Pm = 10^{15} m

pi [*py*] = π, 16th letter of the Greek alphabet. The ratio of the circumference (c) of any circle to its diameter (d) $\pi = c/d$.

Pi cannot be expressed exactly as the ratio of two whole numbers, nor does it occur as the solution to any algebraic equation. It is an example of an **irrational** number and a **transcendental** number.

An approximate value of π is 3.14159. Most electronic calculators have a special key for π.

Pick's rule   A rule for finding the area of a **polygon** drawn on dot paper.

The area (A) of a polygon is given by the formula
$$A = \tfrac{p}{2} + n - 1,$$

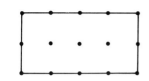

Figure P18

where *p* is the number of dots on the **perimeter** and *n* is the number of dots inside the polygon.
Example
$p = 12, n = 3$
$A = \frac{12}{2} + 3 - 1 = 8$, so area is 8 cm².
For some less regular shapes, the formula gives approximate answers only.

pico- [*pik*-oh] A prefix meaning one-million-millionth. Symbol p.
Example
1 picometre = 0.000 000 000 001 metre
1 pm = 10^{-12} m
SPANISH *pico*: sharp point

pie chart A way of showing how some quantity is divided up. It consists of a circle marked off in **sectors**, each sector representing a share of the whole. The angles of the sectors (and hence their areas) are proportional to the shares.
Example

Mode of transport	No. of students	% of students
walk	288	40
bicycle	144	20
bus	180	25
car	108	15
	720	100

Figure P19

The results are represented graphically in the pie chart shown in Figure P19.
 Compared with a **bar chart** or **column graph**, a pie chart does not show up small differences very clearly.

place value The basic idea of systems of **notation** like the **decimal** number system in which the value of a **digit** depends on where it is placed in a string of digits.

Example
In the following examples of decimal notation, the digit 2 has the different values: two, twenty, two hundred and two-hundredths respectively: 542; 528; 3204; 84.52. Places next to each other differ by a factor of ten.

In **binary** notation, adjacent places differ by a factor of two, and in **octal** notation the factor is eight. **Zero** (0) is used to mark a place not otherwise filled.

In contrast to the above examples, **Roman numerals** are not based on a well developed system of place value.
See also **Arabic numerals**.

plan   A drawing to scale of a solid object as seen from above.
See also   **elevation**.
LATIN   *planus*: flat

plane   A flat surface. It has the property that the line joining any two points in the surface lies wholly within the surface. One and only one plane exists through any three points not in the same straight line. **Circle, triangle, polygons** are examples of plane figures; they are two-dimensional.

The word is also used as an adjective.
LATIN   *planus*: flat

plane figure   A geometric figure that lies wholly in one plane, e.g. circles, ellipses, polygons are examples of plane figures. Cubes and other **polyhedra** are not plane figures, even though all their faces are plane.

plane geometry   That part of the study of **geometry** that is concerned only with figures all drawn in the one plane. It includes one-dimensional and two-dimensional geometry, but not three-dimensional (solid) geometry.

plane table A horizontal table used in making a map of an area by taking sightings of landmarks.

A base line on the ground is represented to scale by a line segment on the plane table (see Figure P20). With the plane table first at one end and then the other end of the base line, lines of sight are drawn to locate points on the plane table that correspond to the landmarks. In this way a scale map of the area is built up.

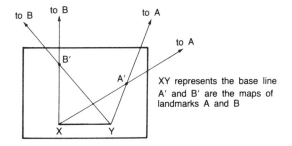

XY represents the base line
A' and B' are the maps of landmarks A and B

Figure P20

planimeter [plan-*im*-e-ter]   An instrument that measures the area of a plane figure. Part of the instrument is a pointer, which is moved around the **perimeter** of the figure, and the area is read directly from a dial.

Plato [*play*-toh]   A Greek scholar who lived in the 4th century BC.

Some modern mathematicians adopt the view, derived from Plato's philosophy, that numbers and other mathematical objects have an independent existence and are not mere inventions of the mind.

Platonic solids   The five kinds of regular **polyhedron**: cube, regular tetrahedron, regular octahedron, regular dodecahedron, regular icosahedron. These were proved in the time of

Plato to be the only possible regular polyhedra. *See* **regular**.

platykurtic (a) In statistics, referring to a distribution that is less concentrated around the mean than a **normal distribution**. The graph is flatter than a **normal curve**.
Contrast **leptokurtic** and *see* **kurtosis**.
GREEK *platys*: flat, *kurtosis*: curvature

plot (v) To mark the position of points relative to a system of **coordinates**; to draw a graph through these points.
Example
To plot the graph of $y = 1/x$, a table may be drawn up of values of x and corresponding values of y:

x	-4	-2	-1	$-\frac{1}{2}$	$-\frac{1}{4}$	$\frac{1}{4}$	$\frac{1}{2}$	1	2	4
y	$-\frac{1}{4}$	$-\frac{1}{2}$	-1	-2	-4	4	2	1	$\frac{1}{2}$	$\frac{1}{4}$

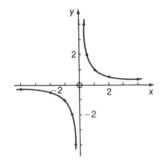

Figure P21

Then the points and curve are plotted as in Figure P21.

plus sign The symbol + used to indicate **addition**, e.g. $5 + 3 = 8$, read 'five plus three equals eight'.
Also called 'positive sign' and used to distinguish a **positive** number or the positive sense of a direction from a negative number or a negative sense, e.g. $+2$, read 'positive two' or 'plus two'.
See also **directed number**.

point 1 A basic element in geometry, marking position but having no size or area. Lines, planes and geometric figures consist of sets of points. In drawn diagrams, the position of a point is often marked approximately by a visible dot. 2 The name given to the dot used in systems of notation like the decimal system,

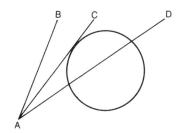

Figure P22

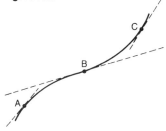

Figure P23

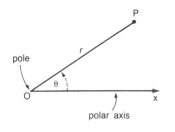

Figure P24

e.g. 2.51, read 'two point five one'.
LATIN *punctum*: puncture, point

point of contact A **point** at which two geometric figures touch without intersecting.
Example
In Figure P22, AC shares a point of contact with the circle, but AB and AD do not.

point of inflection A **point** on a curve at which the **tangent** to the curve intersects the curve.
Example
In Figure P23, B is a point of inflection; A and C are not. On moving along the curve, the sense of rotation changes at the point of inflection from clockwise to anticlockwise or vice versa.

polar coordinates A way of defining the position of a point in a plane by (1) its distance from a fixed point (the pole), and (2) the angle of rotation of the line joining the point to the pole from a fixed ray (the polar axis), which has the pole as its starting point.
Example
In Figure P24, O is the pole and OX is the polar axis. The point P has the polar coordinates (r,θ). r is the length of OP and θ is the angle XOP.
Polar coordinates may be converted to **Cartesian** coordinates (x,y) as follows: $x = r \cos \theta$, $y = r \sin \theta$.

polygon A closed plane figure with straight sides. The simplest polygon is a **triangle**. Some other polygons are **quadrilateral** (four sides), **pentagon** (five sides), **hexagon** (six sides), **heptagon** (seven sides), **octagon** (eight sides).
A regular polygon is one that has all its angles equal and all its sides equal.
A **concave** polygon has at least one interior angle greater than two right angles. Other polygons are called **convex** polygons. The

interior angles of a convex polygon with n sides add up to $(2n - 4)$ right angles.

GREEK *polys*: many, *gonia*: an angle

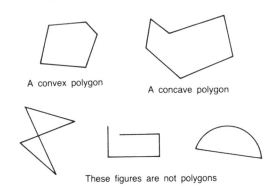

Figure P25

polygonal numbers [po-*lig*-on-l] Numbers named after the **polygons**, such as triangular numbers, square numbers, pentagonal numbers. For a description, *see* **figurate numbers**.

polyhedron [pol-ee-*hee*-dron], (pl. **polyhedrons** or **polyhedra**) A solid with plane faces. It follows that each face must have straight sides (must be a **polygon**).

A regular polyhedron has all faces the same, with all pairs of adjacent faces making equal angles with each other. There are only five kinds of regular polyhedron: **cube, tetrahedron, octahedron, dodecahedron** and **icosahedron**.

Figure P26 shows two examples of a five-faced polyhedron (a **pentahedron**); it cannot be regular.

 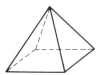

Figure P26

Polyhedrons are named after the number of faces, e.g. (5) penta-, (6) hexa-, (7) hepta-, (8) octa-, (10) deca-, (12) dodeca- (20) icosa-hedron.
GREEK *polys*: many, *hedron*: a base

polynomial [pol-i-*noh*-mi-al] An algebraic expression consisting of two or more **terms**. It is usual for the expression to contain only one **variable** (or unknown) and for this variable to be raised to non-negative whole number powers only.
Example
$x^3 + 2x^2 + x - 1$ is a polynomial (here x is raised to the powers 3,2,1,0); but $x^2 + 2/x + 1$ is not a polynomial (the second term has x raised to the power -1) and $2x^2 - \sqrt{x}$ is not (the second term has x raised to the power $\frac{1}{2}$).

The simplest cases of a polynomial are **binomial** (two terms) and **trinomial** (three terms).
GREEK *polys*: many, *nomen*: a part or name

population In statistics, the total set of persons, things or events under investigation. If a population is large, a **sample** may be drawn from it and the characteristics of the population then inferred from the characteristics of the sample, using the techniques of inferential **statistics**.
LATIN *populus*: people

positive (a)
1 Of a **directed number**, one that is greater than zero. On the **number line**, positive numbers are usually to the right of zero while **negative** numbers are to the left. A positive number may be written without a sign or with a **plus sign**, e.g. positive 3 (or plus 3) is written 3 or $^+3$ or $+3$.
2 Of the sense of a direction. Opposite senses of a direction are labelled positive and negative.

Positive angles are measured as an anticlockwise rotation, to distinguish them from negative angles, which are measured in a clockwise direction.

A positive acceleration is an acceleration in the same sense as the motion of the body; that is, it is a speeding up, as distinct from a negative acceleration, which is a slowing down. In a problem about falling objects, downwards may be chosen as positive and upwards as negative, or the other way about.

LATIN *positivus*: settled by agreement

post meridiem (adverb) [*pohst* mer-*id*-ee-em] Abbreviation: p.m. Used in quoting time of day as given by a 12-hour clock, e.g. 9 p.m. means nine o'clock in the evening.
Compare **ante meridiem**.
LATIN *post*: after, *meridien*: midday

power The result obtained by multiplying a number by itself one or more times.
Example
16 is the fourth power of 2, since $2 \times 2 \times 2 \times 2 = 16$. 'The fourth power of 2' may also be read as '2 to the power 4'. It is written 2^4.

The second and third powers of a number are given special names drawn from geometry: n^2 is the square of n or 'n squared', n^3 is the cube of n or 'n cubed'.

In advanced mathematics, a meaning is given to powers that are not natural numbers.
Examples
a 2 to the power -1 (written 2^{-1}) is $\frac{1}{2}$
b 5 to the power $\frac{1}{2}$ (written $5^{\frac{1}{2}}$) is $\sqrt{5}$
c 3 to the power 0 (written 3^0) is 1
LATIN *posse*: be able

premise [*prem*-is] (Also spelt **premiss**) A statement assumed to be true, and from which a conclusion is drawn, e.g. premise — p is a prime number greater than 2; conclusion — p must be an odd number.
LATIN *praemissa*: sent before

prime A mark (′) used to distinguish one letter or symbol from another with the same name, e.g. after translation the new position of triangle ABC may be labelled A′B′C′.
See also **prime number**.

prime number or **prime** A natural number having no **factors** except itself and one. The first few primes are 2,3,5,7,11,13. It is not usual to include 1 among the prime numbers.
Every natural number has its own single way of being expressed as a product of primes, e.g. $36 = 2 \times 2 \times 3 \times 3$; $130 = 2 \times 5 \times 13$. This is why primes are seen as the building blocks of the natural numbers, and have interested many people for centuries. A natural number that is not a prime is a **composite** number.
LATIN *primus*: first

prism A solid having two **polygon**-shaped faces that are parallel and congruent and other faces that are **parallelograms**. This means also that all **cross-sections** parallel to the two parallel faces are the same shape and size.
Example
See Figure P27.
GREEK *prisma*: something sawn

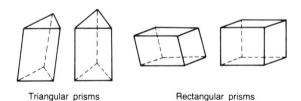

Triangular prisms   Rectangular prisms

Figure P27

probability The confidence that some event will happen, measured or estimated on a scale of 0 to 1. Zero probability means impossibility, a probability of 1 means certainty, a probability of $\frac{1}{2}$ is sometimes called a fifty-fifty chance or an even chance.

The mathematical probability of a particular event occurring as the result of an experiment is the **ratio** of the number of ways the event can occur to the total possible number of outcomes of the experiment.

Example
If two dice are thrown, there are thirty-six possible ways they can land, and of these ways six give a score of 7 (1-6, 2-5, 3-4, 4-3, 5-2, 6-1). The probability of throwing a 7 is therefore calculated as 6/36 = 1/6. The probability of not throwing a 7 is 5/6. These two probabilities add to 1 (certainty).

The probability of a particular event occurring in a **population** may be estimated by observing the actual number of such occurrences in a **sample** and expressing this as a ratio of all outcomes.

See also **law of large numbers**.

LATIN *probabilis*: likely

problem A question put forward for solution or discussion.
1 A mathematical question.
Examples
a What is the value of x in the equation $2x + 3 = 13$?
b How many **prime** pairs like (11,13), (17,19), (59,61) are there?
2 Mathematics may be used in attempting to solve problems that are not purely mathematical, *see* **modelling**.

GREEK *pro-*: forward, *blema*: thrown

produce (v) [pro-*dyoos*] To extend a **line segment**, such as the side of a geometrical figure, e.g. in Figure P28, two sides of the trapezoid have been produced to meet at the point P.

LATIN *producere*: to bring forward

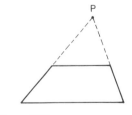
Figure P28

product The result of **multiplication**, e.g. the product of 5 and 12 is 5 × 12 = 60. In algebra, the product of numbers a and b is written as $a \times b$ or $a \cdot b$ or ab.
Compare **sum**.

program A sequence of instructions (an **algorithm**) for a computer to follow in completing a task.
GREEK *programma*: a public notice

progression A **sequence** of numbers in which there is a constant relation between each number and the following one, e.g. 1,2,4,8,..., each number being half of the following one.
See **arithmetic progression, geometric progression, harmonic progression, Fibonacci**.
LATIN *progredi*: to go forward

projection A **mapping** of a geometric figure on to a line or plane (rather like the formation of a shadow).
Example
In Figure P29, the projection of line segment PQ on to the x-axis is AB and on to the y-axis is CD.
Projections are important in map-making where parts of the earth's surface (curved) are to be represented on plane paper. Consult a good atlas for a description of the various ways of doing this.
LATIN *pro-*: forward, *jacere*: to throw

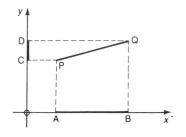

Figure P29

pronumeral A symbol standing in place of a **numeral**.
Example
If it is known that there are twenty-four students in a class, this number can be represented by the numeral 24. But, if the size of the class is not known or if it is variable, it is represented by a pronumeral such as N. A pronumeral can be a letter or any other symbol that would not be

confused with an ordinary numeral. Pronumerals are important in algebra because they allow general mathematical statements to be made e.g. $a(a + b) = ab + ac$, whereas in arithmetic only particular statements are made, e.g. $2(3 + 4) = 2 \times 3 + 2 \times 4$.

LATIN *pro-*: in place of

proof A procedure for showing a proposition to be true by arguing logically from **axioms** or other propositions assumed or known to be true. This method of proof is an example of deductive reasoning. It is different from the inductive method of reasoning used in science, where theories are proved by appeal to observation and experiment. An important proposition of the kind that can be proved mathematically is called a **theorem**.

proper fraction A **fraction** that is expressed with a numerator and denominator, the numerator being smaller than the denominator, e.g. $\frac{1}{2}$ and $\frac{3}{4}$ are proper fractions, whereas $\frac{5}{2}$ is **an improper fraction**. A proper fraction always has a value less than 1.
Also called simple fraction.

proportion Two pairs of numbers are in proportion if the **ratio** formed by the first pair equals the ratio formed by the second pair.
Example
10 and 5 are in proportion to 16 and 8, since 10/5 and 16/8 have the same value (2). This can be written 10:5::16:8 read '10 is to 5 as 16 is to 8'.
In geometry, the lengths of the sides of one figure are in proportion to those of a **similar** figure, e.g. if ABC and DEF are similar triangles, then $AB/DE = BC/EF = CA/FD$ (see Figure P30).

Two pairs of numbers are in **inverse proportion** if their ratios are the **reciprocals** or **inverses** of each other, e.g. 10/5 and 8/16 are in

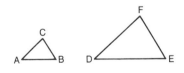

Figure P30

inverse proportion, since 10/5 = 2/1 and 8/16 = 1/2.
See also **direct proportion, variation**.

proportional (a)   In **proportion**. One variable is proportional to another if the **ratio** of corresponding values remains constant.
Example
In the following table, the values of x and y are in proportion, suggesting that y/x is always equal to the same value (5). If this is so, then y is proportional to x, written $y \propto x$.

x	1	2	3	4	5
y	5	10	15	20	25

proposition
1 The formal statement of a theorem or problem.
2 In logic, a sentence that is capable of being either true or false, but not both.
  A proposition is often represented by a lower case letter such as p or q.
Examples
The following sentences are propositions, because we know they must be true or false:
a 'Canberra is the capital of Australia.'
b '10 is exactly divisible by 3.'
The following sentence is not a proposition, as there is no way of deciding whether it is true or not: 'This number is divisible by 5.'
LATIN   *proponere*: to put forward

protractor   An instrument for measuring angles. It is usually in the shape of a circle or semicircle and is marked around the edge in degrees.
LATIN   *protrahere*: to draw out

Figure P31

pyramid [*pir*-a-mid]   A solid with any **polygon** for a base, the other faces being triangles meeting at a point (**vertex**). A right pyramid has a regular base, and its vertex lies on a line drawn

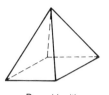

Pyramid with square base

Pyramid with hexagonal base

Figure P32

perpendicular to the centre of the base.
The **volume**, V cm^3, of a pyramid is calculated from the formula, $V = \frac{1}{3}Ah$, where A square centimetres is the area of the base and h centimetres is the perpendicular distance of the vertex above the base.

LATIN *pyramis*, from Greek and perhaps Egyptian origin

Pythagoras [py-*thag*-or-as] A 6th century BC Greek mathematician, philosopher and mystic. In mathematics, he and his followers are credited with the proof of the theorem that bears his name (*see* next item), with the discovery that the ratio of the length of the diagonal to the length of the side of a square cannot be expressed as a rational number, and with the properties of **figurate numbers**.

Pythagoras's theorem For a right-angled triangle, the square on the longest side (the hypotenuse) is equal to the sum of the squares on the other two sides.
The **converse** is also true.
The theorem can be demonstrated by dissecting and rearranging a cardboard cut-out as suggested in Figure P33. The smallest square is placed in the largest square and surrounded by four pieces cut from the remaining square. The cut lines for these pieces pass through the centre

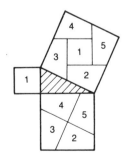

Figure P33

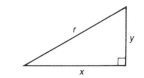

Figure P34

Figure P35

of the square and are parallel to the sides of the largest square.

The result of Pythagoras's theorem is written algebraically as $x^2 + y^2 = r^2$, where r is the length of the hypotenuse and x, y are the lengths of the other two sides as in Figure P34.

An important special case is that the length of the diagonal of a unit square is $\sqrt{2}$ (see Figure P35).

Pythagorean triple [py-thag-or-*ee*-an *trip*-1] Expressed geometrically, any set of three whole numbers that can represent the lengths of the sides of a right-angled triangle. Expressed algebraically, any set of three whole numbers that satisfies the equation $a^2 + b^2 = c^2$.

Examples
a (3,4,5), since $3^2 + 4^2 = 5^2$.
b (5,12,13).
c (8,15,17).

Q

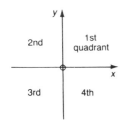

Figure Q1

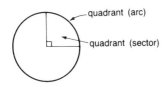

Figure Q2

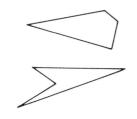

Figure Q3

quadrant [*kwod*-rant] **1** One of the four plane regions marked out by a pair of coordinate axes, as in Figure Q1. **2** A quarter of a circle, defined as either a 90° **arc** or the corresponding **sector** (Figure Q2).
LATIN *quadrans*: quarter

quadratic (a) [kwod-*rat*-ik] Referring to an algebraic expression, function or equation that involves the second **power** (square) as the highest power of the unknown.
Examples
a $2x^2 + 3x - 1$ is a quadratic expression.
b $y = 3x^2$ expresses y as a quadratic function of x.
c $3x^2 - 5x - 2 = 0$ is a quadratic equation.
The general form of a quadratic expression is $ax^2 + bx + c$.
Sometimes, a quadratic equation or function is referred to simply as a quadratic.
LATIN *quadratus*: made square

quadrilateral [kwod-ri-*lat*-ral] Any plane figure having four straight sides (see Figure Q3). Special quadrilaterals include **kite, chevron, parallelogram, rectangle, rhombus, square**.
LATIN *quadri-*: four, *latus*: a side

quadrillion [kwod-*ril*-i-on] In North American usage, 1 quadrillion = 1 thousand million million, i.e. 1 000 000 000 000 000 or 10^{15}.
In British usage, 1 quadrillion = 1 million million million million, i.e. 1 000 000 000 000 000 000 000 000 or 10^{24}.
LATIN **quadri-**: four + (m)illion

quantity [*kwon*-ti-tee] Any property of an object whose size or magnitude can be expressed by a number.
Examples
The following are quantities: the number of elements in a set; the length of a line segment; the capacity of a jug; the velocity of a projectile; the size of a bank balance.
See also **scalar, vector**.

quartile [*kwor*-tyl] In a **frequency distribution**, any one of the three values of the variable that divides the total distribution into four parts of equal frequency. The three values lie at the lower quartile, the **median**, and the upper quartile. The lower quartile separates the first 25% of the distribution from the 75% remaining. The upper quartile separates the first 75% from the 25% remaining. The median divides the distribution equally.
Example
If 25% of the population are no taller than 105 centimetres, then 105 centimetres is the height value lying at the lower quartile.
See also **percentile**.
LATIN *quartus*: fourth

Figure Q4

quincunx [*kwin*-kunks] An arrangement of five objects with one at each vertex of a square and one in the centre, e.g. the number five on a die (Figure Q4).
LATIN *quinque*: five, *uncia*: twelfth part. Originally a Roman coin worth 5/12 of an *as*.

quotient [*kwoh*-shent] The result of dividing one number by another, e.g. if 12 is divided by 5, the quotient is 2.4.
LATIN *quotiens*: how many times

R

radial (a) [*ray*-di-al] Moving in a straight line from a point; like a **ray** or like the **radius** of a circle.

radian A unit of angle measurement. An angle of 1 radian lies between two radii of a circle that cut off between them an arc on the circumference equal in length to the radius. Symbol: rad.

Because the circumference of a circle is 2π times its radius, there are 2π radians in a full circle:
- 2π rad = $360°$
- π rad = $180°$
- $\frac{1}{2}\pi$ rad = $90°$
- 1 rad = $57.3°$ approximately.

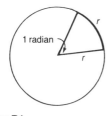

Figure R1

radical [*rad*-i-kal] **1** Another name for the **root** of a number. Square root, cube root, etc. are radicals. **2** The name of the sign for a root, $\sqrt{}$ or $\sqrt[n]{}$ as in $\sqrt{5}, \sqrt[3]{27}, \sqrt{a+b}$.
LATIN *radix*: a root

radius A line segment joining the centre of a circle or sphere to a point on the circumference; or the length of this line segment.
LATIN *radius*: spoke of a wheel

radix [*ray*-diks] A number used as a **base** for a number system.
Example
The decimal system is based on ten **digits**, so its radix is ten. The binary system is based on two digits, so the base is two. Common **logarithms** have the base or radix ten.
LATIN *radix*: a root

raise (v) To multiply a number by itself to a given number of factors, e.g. to raise 2 to the power 3 means to calculate $2^3 = 2 \times 2 \times 2 = 8$.

random (a) Not following a pattern or rule. Occurring by chance.

random numbers table A list of numbers arranged in chance order so that no number can be predicted from preceding ones. Computer programs have been written to generate random numbers. Random number tables are used in statistics to help draw **random samples** from a population.

random sample A **sample** taken from a **population** (or set) in such a way that it has the same characteristics as the population (or set). A random sample can therefore be accepted as representative of the population (or set) as a whole. The technique of taking a random sample is important in statistics, as for example in opinion polls.

random variable A quantity that has a range of possible values, none of which can be predicted with certainty.
Example
If two dice are thrown, the score has a range of possible values: 2,3,4,5,6,7,8,9,10,11,12, but the actual outcome of one throw is a random variable and can be predicted only as a **probability**. The probability of throwing a score of 8, for instance, is 5/36, because there are five different ways of making up a score of 8, and 36 ways altogether in which the two dice can fall.

range
1 The set of all values adopted by a **function** as the independent variable takes on all the values of the **domain**.

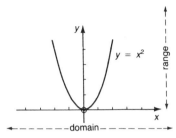

Figure R2

Example
$y = x^2$. The set of values available to x is the range. In this example, the domain is all values of x from $-\infty$ to $+\infty$, but the range is restricted to zero and all positive values of y (see Figure R2).

2 In statistics, the difference between the largest and smallest values of a variable in the sample under investigation.

Example
The heights of students in a class are measured. It is found that the shortest student is 140 centimetres tall and the tallest is 205 centimetres. The range then is $205 - 140 = 65$ centimetres.

rank (v) To arrange a set in order of some measure, e.g. the following people are ranked in order of height: Tom (162 cm), Lisa (165 cm), Joe (170 cm).

The arrangement is called rank order, and may be either ascending (from lower to higher) or descending (from higher to lower).

rate One quantity measured in relation to another quantity, e.g. the speed of an object is the rate at which its position changes with time, measured in kilometres per hour, etc.

In a rate, the two measures being compared are expressed in different units, whereas in a **ratio** the units must be the same.

rate of increase The **gradient** of a **function**. For a function $y = f(x)$, this means the rate at which y increases with respect to x.

For y with respect to x, the following figures show:
- a zero rate of increase (Figure R3)
- a constant rate of increase (positive) (Figure R4)
- an increasing rate of increase (Figure R5)
- a decreasing rate of increase (Figure R6)

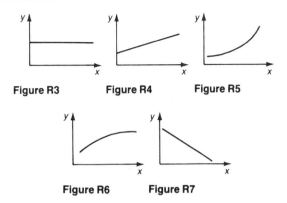

Figure R3   Figure R4   Figure R5

Figure R6   Figure R7

- a negative constant rate of increase (a constant rate of decrease) (Figure R7).

ratio A number or quantity compared with another, and expressed with the symbol,:, or written as a **fraction**.
Example
The ratio of 12 to 24 is 12:24 or 12/24 or $\frac{1}{2}$. The ratio of the circumference of a circle to its diameter is $c:d$ or c/d ($= \pi$). If an interest rate of 10% is increased in the ratio of 1:1.5, then the new interest rate will be $10 \times 1.5 = 15\%$.
See also **scale**.
LATIN   *ratio*: calculation

rational number A number that can be expressed exactly as the **ratio** of two **integers**. If a and b are integers, then a/b is a rational number if b is not equal to zero. Examples of rational numbers are $\frac{1}{8}$, 14/2, $-5/2$, $3\frac{1}{2}$ ($=7/2$). All terminating and recurring decimals are rational, e.g. $1.414 = 1414/1000$, $0.33333\ldots = \frac{1}{3}$.

Numbers that cannot be expressed as the ratio of integers are called **irrational**, e.g. $\sqrt{2}, \sqrt{5}, \pi, e$.

The set of rational numbers is **closed** under all the ordinary operations of addition, subtraction, multiplication and division.
See also **number**.
LATIN   *ratio*: calculation

rational function A **polynomial**, or the **ratio** of polynomials. This means that a rational function of x involves only whole number **powers** of x and all the **coefficients** are rational numbers.
Example
$x^2 + 1$; $(3x^3 - 1)/(x + 1)$ are rational functions of x.
$\sqrt{x - 1}$ is not a rational function.

raw data Statistical information as it is collected, and before it has been grouped or adjusted in any way.
Examples
The number and kind of vehicles passing a busy intersection in a traffic survey; a list of test results for a mathematics class.

ray A half **line**.
Example
In Figure R8, AP and AQ are rays starting at point A and extending infinitely in the directions and senses of the arrows.
Any single line can be divided into two rays as in Figure R9.
LATIN *radius*: spoke of a wheel

Figure R8

Figure R9

real number A number that is either **rational** or **irrational**. Real numbers include positive and negative **integers, zero, fractions**, and irrational numbers like $\sqrt{2}$ and π. They do not include **imaginary** or **complex** numbers.
Real numbers may be represented on a **number line**.
See also **number**.

reciprocal [re-*sip*-ro-kal] The multiplier of a number that gives 1 as the result.

Example
½ is the reciprocal of 2, and 2 is the reciprocal of ½, since ½ × 2 = 1. Some other pairs of reciprocals are: 3/4 and 4/3; 1.5 and 2/3; a/b and b/a; x^2 and x^{-2}.
Also called multiplicative inverse.
LATIN   *reciprocus*: turning back

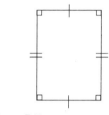
Figure R10

rectangle   A right-angled **parallelogram**. A rectangle has all the properties of a parallelogram (opposite sides parallel and equal, etc.) with the additional property that its angles are right angles and its diagonals are equal, as in Figure R10.
A **square** is a special case of a rectangle having all sides equal.
LATIN   *rectus*: right + angle

rectangular number   Any positive **integer** that is not a **prime number**. It can therefore be expressed as the product of two smaller integers and can be represented as a rectangular array.
Example
12 can be expressed as 4 × 3 and pictured as a 4 × 3 array of dots (Figure R11).
See also   **figurate numbers**.

Figure R11

rectilinear (a) [rek-ti-*lin*-i-ar]   Formed from straight lines, or acting in a straight line. All **polygons** are rectilinear figures; **circle** and **ellipse** are not.
Rectilinear motion is the movement of an object travelling in one direction only.
Also rectilineal.
LATIN   *recti-*: straight, *linea*: line

recurring (a)   Of a decimal fraction that contains an endlessly repeating digit or pattern of digits.
Examples
a   1/9 = 0.1111..., written 0.1̇ and read 'point one recurring'.

b 4/30 = 0.1333..., written 0.1$\dot{3}$, read 'point one, three repeating'.
c 1/7 = 0.142857142857..., written 0.$\dot{1}$4285$\dot{7}$. Also known as repeating.
LATIN *re-*: back, *currere*: run

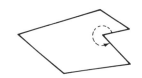

Figure R12

re-entrant (a) Describing a **polygon** having at least one interior angle greater than 180°, e.g. a re-entrant hexagon as in Figure R12.

reference point A point acting as a marker or **origin** from which the positions of other points are described, e.g. the origin in a **Cartesian** graph is a reference point for displacements measured along both axes.

reflection A **transformation** of a point, line or figure that results in a mirror image of it. Reflection always takes place in a straight line, and this line acts as a line of **symmetry** between the original object and its image (see Figure R13).
See also **flip**.
LATIN *re-*: back, *flectere*: bend

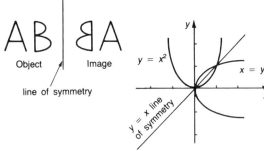

Figure R13

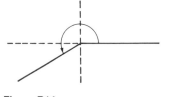

Figure R14

reflex angle An angle that is greater than two right angles and less than four right angles (Figure R14). It measures between 180 and 360 degrees, and between π and 2π radians.
LATIN *re-*: back, *flectere*: bend

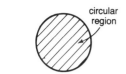

Figure R15

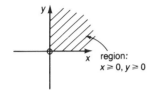

Figure R16

region A part of a **plane**. It may be enclosed, e.g. the interior of a circle (Figure R15), or it may be only partly bounded, e.g. the first **quadrant** of a **Cartesian** plane (Figure R16).

In this definition, a region is **two-dimensional**, but the term is sometimes extended to refer to a part of **three-dimensional** space, e.g. the interior of a sphere.

LATIN *regio*: a boundary line

regular (a) **1** Of a **polygon**, having equal sides and equal angles. **2** Of a **polyhedron**, having identical polygon-shaped faces and equal angles between adjacent faces.

Any regular polygon can have both an **incircle** and a **circumcircle**.

Any regular polyhedron can have an inscribed sphere and a circumscribed sphere.

Only five kinds of regular polyhedra are possible as shown in Figure R17. The first three of these shapes occur in natural crystal formations, e.g. common salt crystals are cubic.

LATIN *regularis*: of a ruler

Tetrahedron (four faces)   Hexahedron (six faces)   Octahedron (eight faces)   Dodecahedron (twelve faces)   Icosahedron (twenty faces)

Figure R17

relation A property connecting two or more sets of numbers or elements.

Examples

a F is the father of S: 'is the father of' is the relation connecting one set of men (S) to another set (F).

b $y = x^2$: 'is the square of' is the relation connecting one set of numbers (x) with another set (y).

A relation can be described as a set of **ordered pairs**:

In **a** above, the set of ordered pairs may include (Nicolas, John), (Paul, John), (Jason, Steve). The second member of each pair is the father of the first member.

In **b** above, some members of the set of ordered pairs are (1,1), (2,4), (−2,4), (3,9). The second member of each pair is the square of the first member. Such a relation can be illustrated by the graph of $y = x^2$.

Where (as in the case of $y = x^2$) each value of x has no more than one value of y associated with it, the relation is a **function**.

The examples given are called binary relations, because they involve two variables (father and son, x and y). They should not be confused with **binary operations**, such as $a + b$ and $a \div b$.

LATIN *relatio*: a throwing back

relative (a) [*rel*-a-tiv]   Expressed as a proportion of a total, as in relative **frequency**.
Example
If the result of tossing a coin 100 times was 53 heads and 47 tails, the relative frequency of heads would be expressed as 0.53 or 53%. Note that the relative frequencies of all the events must add up to 1 (or 100%).

LATIN *relatus*: carried back

remainder   In arithmetic, the amount left over when one number is divided by another, e.g. when 20 is divided by 6, the remainder is 2, since $20 = 6 \times 3 + 2$.

The term is used in a similar way in algebra, e.g. if $x^2 + 3x + 2$ is divided by $x + 2$, the remainder is 0, since $x^2 + 3x + 2 = (x + 2)(x + 1) + 0$.

LATIN *re-*: back, *manere*: to stay

remainder theorem   The theorem that states that, if a **polynomial** F(x) is divided by ($x - a$),

then the **remainder** will equal the number obtained by substituting a for x in $F(x)$.
Example
If the polynomial $2x^3 + x^2 - 7x - 6$ is divided by $(x - 1)$, the theorem tells us that the remainder can be found by substituting $x = 1$, giving $2 + 1 - 7 - 6 = -10$.

The theorem is used to help in finding factors, for dividing by a factor always leaves a zero remainder, e.g. in the above example, substituting $x = 2$ shows the remainder to be 0, and hence $(x - 2)$ must be a factor.

rename (v)   To rewrite a mathematical expression in a different form without changing its value, e.g. $\tfrac{3}{4}$ may be renamed as 6/8 or 9/12, etc.
Renaming is useful in adding fractions, e.g. to find $\tfrac{3}{4} + \tfrac{1}{8}$, rename as 6/8 + 1/8 = 7/8.

representative sample   In statistics, a **sample** drawn from a population in such a way that the properties of the sample accurately reflect the properties of the population. A true representative sample is free from bias.
See also   **random sample**.

resolution   The replacement of a single **vector** by two other vectors, having together an equivalent effect. The two replacement vectors are usually chosen at right angles to each other and are then known as resolved parts or resolutes.
Example
The velocity of a projectile at a given moment may be resolved vertically and horizontally as shown in Figure R18, so that the vector sum $\mathbf{V}_x + \mathbf{V}_y$ is the equivalent of the original $\mathbf{V}$.
LATIN   *resolvere*: to untie
See also   **vector**.

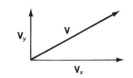
Figure R18

resultant   The single **vector** that is the sum of two or more given vectors.

Figure R19

Example
The forces acting at a point P and represented by vectors F_1 and F_2 as shown in Figure R19 can be replaced by a single force represented by vector R in the manner shown. R is then the resultant of F_2 and F_2.
LATIN *resultare*: to spring back

revolution A complete turn. An object completing one revolution turns through an angle of 360° equivalent to four right angles.
LATIN *revolvere*: to roll back

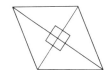

Figure R20

rhombus [*rom*-bus] A **parallelogram** with equal sides. Sometimes called diamond or rhomb. The diagonals of a rhombus are at right angles to each other (see Figure R20).
GREEK *rhombos*: spinning top

right (a) Having a **right angle**.
Example
A right circular **cone** and a right circular **cylinder** each have an axis at right angles to the base. A triangle with a right angle is sometimes called a right triangle.

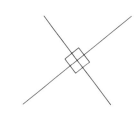

Figure R21

right angle Each of the angles formed when two lines intersect to form four equal angles as in Figure R21. A right angle represents a quarter turn, and its measure is 90 degrees or $\pi/2$ radians.

Roman numerals Numerals in the ancient Roman system of notation. The basic symbols (with their modern decimal equivalents) are: I(1), V(5), X(10), L(50), C(100), D(500), M(1000).

The system is based on counting in tens, so might be described as a decimal one, but it has no symbol for zero and it does not use the idea of **place value** as in the system of **Arabic numerals**.

The main rules for building up numerals from the basic symbols are as follows:
- To lower the value of a symbol, place a lower value symbol before it, e.g. V = 5, IV = 4; L = (50), XL = 40.
- To raise the value of a symbol, place a symbol of the same or lower value after it, e.g. V = 5, VI = 6, C = 100, CC = 200, CCX = 210.

Other examples: II(2), III(3), VII(7), VIII(8), IX(9), MCMXCIV (1994).

root

1 Of a number: the number that, multiplied by itself, results in the given number.

Examples
a 2 is the square root of 4, since $2 \times 2 = 4$
b 2 is the cube root of 8, since $2 \times 2 \times 2 = 8$
c 2 is the fourth root of 16, since $2 \times 2 \times 2 \times 2 = 16$
d The square root of x is written $\sqrt{x}$ or $x^{\frac{1}{2}}$
e The cube root of x is written $\sqrt[3]{x}$ or $x^{\frac{1}{3}}$
f The nth root of x is written $\sqrt[n]{x}$ or $x^{1/n}$

2 Of an equation: that number which, substituted for the unknown, makes the equation a true statement.

Example
$x = 2$ is a root of the equation $x^2 - x - 2 = 0$, since it is true that $2^2 - 2 - 2 = 0$.

The root of an equation is also known as a **solution** to that equation.

rotation

rotation A **transformation** of a geometrical figure in which each point of the figure moves around a fixed point by the same angular amount. The fixed point is known as the centre of rotation. It may be inside or outside the figure.

Figure R22 illustrates a 90 degree rotation of a triangle around a point A inside and a point B outside the triangle.

LATIN   *rota*: a wheel

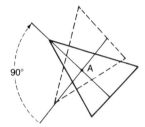
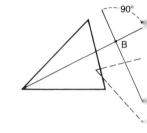

Figure R22

rotational symmetry   If a shape can be turned about a point so that it exactly coincides with its original position at least once in less than a complete rotation, it is said to have rotational symmetry about that point. The point is called the centre of symmetry.
Example
The shape shown in Figure R23 coincides with its original position after turning through 120° around the point O. This shape has rotational symmetry of order 3 because it appears in three identical positions during one complete rotation

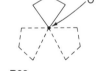

Figure R23

rounding   A rule to follow when making an **approximation** to a given number by using fewer **significant figures**.
Example
When 237 is approximated to the nearest 10, it becomes 240, and when 45.28 is approximated to the nearest 1/10 it becomes 45.3. These are examples of rounding up.

When 234 is approximated to the nearest 10, it becomes 230, and when 45.23 is approximated to the nearest 1/10 it becomes 45.2. These are examples of rounding down.

The rule is to round up when the following digit is 5 or greater, and to round down when the following digit is less than 5. Special care is needed when making successive calculations with rounded numbers to avoid the growth of error.
See also **accuracy**.

row [*roh*]   A horizontal **array** of numbers, as in a **matrix**.
Example
In the matrix shown here, there are two rows : ab and cd.
$$\begin{vmatrix} a & b \\ c & d \end{vmatrix}$$
See also **matrix, column**.

S

sample In statistics, part of a **population**, selected so as to give information about the population as a whole.
Example
The general opinion about some matter of interest to the residents of a city may be gauged by gathering the opinion of a sample made up of one person in every 10 000.
  Ways of drawing a sample in order to avoid **bias** in the conclusions is an important part of the study of **statistics**. In describing a sample, terms such as **mean, median, standard deviation** are used; each of these is referred to as a **statistic** of the sample, to distinguish it from the corresponding **parameter** of the population.
See also **random sample**.
LATIN *exemplum*: example

sample space In statistics, the total set of possible outcomes of an experiment or investigation.
Example
If, in throwing two dice, the variable of interest is the score of each throw, then the sample space consists of $\{2,3,4,5,6,7,8,9,10,11,12\}$. This set is then the **domain** for the **frequency distribution** of the variable.

sampling error If two different **samples** are drawn from the same **population** and then the **mean** and other characteristics calculated, it is likely that differences will be found between the two samples. These differences are the result of sampling error. When only one sample is used from which to infer the characteristics of a population, it is important to be aware of the likelihood of sampling error and to make allowances for it. Ways of making allowances form part of the study of statistics.

satisfy (v) To meet the requirements of an equation, theorem, etc.
Example
$x = 2$ satisfies the equation $3x - 1 = 5$ in the sense that when x is replaced by 2 the equation makes a true statement. $x = 3$ does not satisfy this equation.
   The pair $(x = 1, y = -3)$ satisfies the inequality $5x + y > 0$, but the pair $(1, -5)$ does not.
LATIN   *satis*: enough, *facere*: to do

scalar (a)   Of a quantity, having **magnitude** but no direction property.
Example
Length, area, speed, volume, mass, density, money, temperature are examples of scalar quantities. Their magnitude can be represented on a **number line** or **scale**, and some of them may have negative as well as positive values, but they are not associated with a direction in space.
   Compare with **vector** quantities, which have the property of direction as well as magnitude.
   The word 'scalar' is sometimes used as a noun.
LATIN   *scala*: a ladder

scale
1   A sequence of marks, usually along a line, used in making measurements, e.g. the scales on a thermometer, a ruler, a radio dial.
   A **linear** scale has equal distances representing equal amounts. On a **logarithmic scale**, distances are marked in proportion to the logarithms of the amounts.
2   The **ratio** of a distance measured on a map or drawing of an object to the corresponding distance measured on the object itself, e.g. if a house plan is drawn so that all distances are

reduced to 1/200th of their actual size, the scale is 1:200. This ratio is also called the scale factor.
LATIN   *scala*: a ladder

scalene (a)   Of a triangle, with no equal sides. A scalene triangle may be acute-angled, obtuse-angled or right-angled (Figure S1), but it may not be **equilateral** or **isosceles**.
GREEK   *skalenos*: unequal

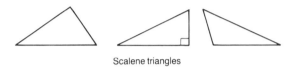

Scalene triangles

Figure S1

scattergram or **scatter plot**   In statistics, a graphical method of showing the joint distribution of two variables, in which each point on the graph represents, by its **coordinates**, a pair of values of the two variables.
Example
Figure S2 shows the test scores in a language and mathematics for a class of twenty students. Each point represents the pair of scores for one student.

The arrangement of points on a scattergram gives a visual idea of the degree of **correlation** between the two variables.

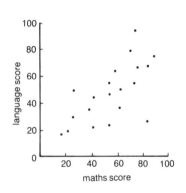

Figure S2

scientific notation   A way of writing a number in the form of a number between 1 and 10 multiplied by a power of 10, e.g. 134.7 and 0.0820 are written 1.347×10^2 and 8.20×10^{-2} respectively.

Note that all digits that appear in scientific notation are to be regarded as **significant figures**.

Some electronic calculators have a key for changing the display between **floating point** and scientific notation.

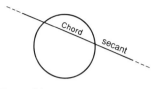

Figure S3

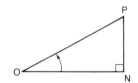

Figure S4

secant [see-kant or sek-ant]
1 A straight line cutting through a circle or other curve. That part of the secant inside the circle is a **chord** (see Figure S3).
2 The **reciprocal** of **cosine**. Abbreviation, sec: sec $\angle NOP = OP/ON$ (Figure S4).
The secant of a given angle is equal to the **cosecant** of the **complementary** angle.
LATIN *secans*: cutting

second 1 An angle measure equalling 1/360 degree. Symbol ″. 2 The standard **SI** unit of time. Symbol s.

section The **intersection** of a **plane** with a surface or solid. It forms a plane figure, e.g. the section formed by a plane cut through a **sphere** is a circle.
See also **conic section, cross-section**.
LATIN *sectio*: a cutting

sector Part of a circle bounded by two radii and an arc between them.
Example
In Figure S5, the shaded region of the circle is a minor sector, the unshaded part is a major sector. The area of the sector $= \frac{1}{2}r^2\theta$, where r is the radius and θ is the **radian** measure of the angle at the centre.
See also **segment**.
LATIN *secere*: to cut

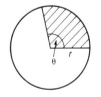

Figure S5

segment
1 Part of a circle cut off by a **chord**. In Figure S6, the shaded region of the circle is a minor segment, the unshaded part is a major segment.
Compare **sector**.
2 Part of a solid figure cut off by one or more planes, e.g. a **frustum** is a segment. 3 *See* **line segment**.
LATIN *segmentum*: a cutting

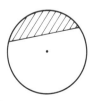

Figure S6

semicircle

Figure S7

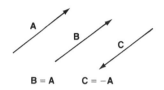

Figure S8

semicircle Part of a circle bounded by a **diameter** and one of the **arcs** joining the end points of the diameter. In Figure S7, both the shaded and unshaded regions are semicircles. They may also be regarded as **sectors** and **segments**. Sometimes it is the half circumference that is called the semicircle.

LATIN *semi–*: half + circle

sense The sign of a direction. A directed line or **vector** has either of two opposite senses, which may be labelled positive or negative. Parallel vectors always have the same direction, but they may have the same or opposite senses (see Figure S8).

LATIN *sentire*: to feel

sequence A set of numbers arranged in order one after the other according to some rule, e.g. 1,3,5,7,... Each of the numbers in the sequence is called a **term**. The nth term in the above example is $2n - 1$.

The general form for a sequence is:
$$a_1, a_2, a_3, \ldots, a_n, \ldots$$
Progressions are examples of a sequence. *See* **arithmetic progression, geometric progression, harmonic progression, Fibonacci**.

series The sum of the terms of a **sequence**.
Example
The sequence 1,3,5,7,... becomes the series $1 + 3 + 5 + 7 + \ldots$ when the terms are added together. The sum of the first four terms of this series is 16; the sum of the first n terms is n^2.

In general, the sum $a_1 + a_2 + a_3 + a_4 + \ldots + a_n$ is written $\sum_{i=1}^{n} a_i$.

See also **arithmetic series, geometric series**.

LATIN *series*: row, chain

set A collection of things defined as members, e.g. the letters of the alphabet; the odd numbers. The individual members of a set are its elements.

A set of things is written with **braces**, e.g. the set of the first five even numbers is {2,4,6,8,10}. This is the same set as {4,2,8,10,6}, as order does not matter.

Capital letters are used to name sets and the Greek letter ϵ **(epsilon)** is used to mean 'is an element of', e.g. A = {4,2,10,8,6,}; 6 ϵ A.

Operations on sets include **intersection** and **union**.

Set theory is the basis for much modern mathematics.

set square An instrument (Figure S9) for help in drawing parallel lines. If it is accurately constructed it may also be used to draw a right angle and other angles (usually 30° and 60° or 45°)

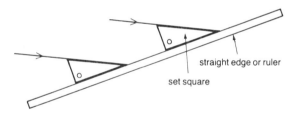

Figure S9

sexagesimal (a) Based on the number sixty. There are traces today of a number system based on sixty as probably used by the ancient Babylonians in the first and second millennia BC, e.g. in angle measurement: 60 seconds in a minute, 60 minutes in a degree, and 6 × 60 degrees in a circle. In time measurement: 60 seconds in a minute, 60 minutes in an hour, and approximately 6 × 60 days in a year.

LATIN *sexagesimus*: sixtieth

SI Abbreviation for **Système International d'Unités** (= International System of Units).

side 1 Any one of the **line segments** forming the boundary of a **polygon**. 2 Any one of the **faces** of a **polyhedron**.

sigma The eighteenth letter of the Greek alphabet, written Σ and σ, corresponding to the English S and s.

Σ is called the summation sign and is used for the sum of terms of a **series** as follows:

$$\sum_{i=1}^{3} a_i = a_1 + a_2 + a_3$$

σ is used in statistics for the **standard deviation** of a **population**.

signed numbers Numbers that are either positive or negative. Positive numbers share the same sign (+), negative numbers share the same sign (−), but a positive number and negative number have opposite signs. Zero has no sign. *See also* **directed number**.

significant figures [sig–*nif*–i–kant] Those digits in a numeral that are relevant to the value of the number represented; the number of such digits is quoted to indicate an approximation.
Example
0.00287 expressed to two significant figures is 0.0029. The first three zeros serve to indicate place value and have no significance apart from this. 0.0029 can be written 2.9×10^{-3} to avoid the use of these zeros.

54.3916 expressed to five significant figures is 54.392, and to four significant figures is 54.40. The zero here is significant as it serves to indicate that the number is greater than 54.39 and less than 54.41.
See also **accuracy**.

similar Of geometrical figures, having the same shape but not the same size. Similar **polygons** have corresponding angles equal and

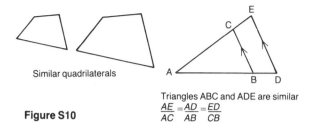

Similar quadrilaterals

Figure S10

Triangles ABC and ADE are similar

$\dfrac{AE}{AC} = \dfrac{AD}{AB} = \dfrac{ED}{CB}$

corresponding sides **proportional**, as shown in Figure S10.

LATIN *similis*: like

simple harmonic motion (SHM) The motion of a particle moving in a straight line, which has an acceleration always directed towards a fixed point in the line (the origin) and which is proportional to the displacement of the particle from that point. The time-displacement graph of such a particle has the shape of a **sine** curve, as shown in Figure S11.

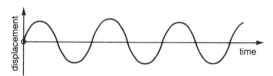

Figure S11

Example
The periodic motion of a weight moving up and down at the end of a coiled spring is SHM. Many periodic motions in nature, science and engineering can be analysed in terms of simple harmonic motion, e.g. the motion of air particles in a sound wave. **Calculus** is used to study them mathematically.

simple interest Interest calculated only on the original amount and not, as in **compound interest**, on the amount to which previous interest has been added, e.g. the simple interest on $100 at 10% per annum is $10 for 1 year, $20 for 2 years, $30 for 3 years, and so on.

simplify (a) To make less complex. In algebra, the action of simplifying produces results like the following:
$$a + 2a + b + 3b = 3a + 4b$$
$$\frac{x^2 - 9}{x + 3} = \frac{(x + 3)(x - 3)}{x + 3} = x - 3$$

simulation The method of studying a complicated system by first creating a mathematical model to represent it and then undertaking calculations on the model, often with the help of a computer. Long-term effects of various factors on the earth's climate may be studied in this way.
See also **modelling**.
LATIN *similare*: to copy, represent

simultaneous equations A set of independent **equations** for which the number of unknowns equals the number of equations, and hence for which there is a common **solution**.
Example
$2x + y = 5$ and $3x - 2y = 4$ are a pair of simultaneous **linear** equations in two unknowns. Their common solution is $x = 2, y = 1$. Note that each equation taken separately is satisfied by many solutions. The first equation, for instance, has solutions $(-1,7), (0,5), (1,3)$ and so on, but there is only one solution shared with the second equation. These facts can be illustrated by a graph, as shown in Figure S12.
LATIN *simul*: at the same time

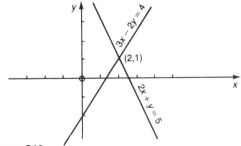

Figure S12

sine curve

sin [*syn*] Abbreviation and symbol for **sine**.

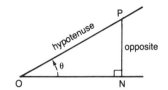

Figure S13

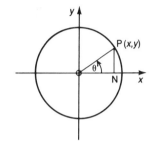

Figure S14

sine Abbreviation and symbol: sin.
1 (trigonometry) — One of the **trigonometric ratios** associated with an angle (Figure S13): sin θ = *PN/OP*.
2 (coordinate geometry) — One of the **circular functions**.
   The sine of the angle θ in the circle shown in Figure S14 is the ratio of the **projection** of OP on the *y*-axis (*y*) to the radius of the circle (*r*): sin A = *y/r*.
   This value is positive when P is above the *x*-axis and negative when below. As P moves around the circle in either direction, θ takes on any value from −∞ to +∞. At the same time, sin θ will fluctuate in value between −1 and +1. It is an example of a **periodic function** (see Figure S15).
See also **cosine, tangent.**
LATIN *sinus*: curve

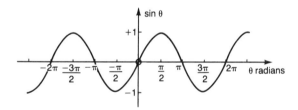

Figure S15

sine curve The graph of sin θ. See Figure S15, where the sine of an angle of θ radians is plotted against θ. Note the following properties:
- The range of values is from −1 to +1. The maximum value (+1) for the sine occurs when θ = π/2 and at intervals of 2π from this. The minimum value (−1) for the sine occurs when θ = 3π/2 and at intervals of 2π from this. The sine is zero for θ = 0 and at intervals of π.

- The **period** of the graph is 2π. The **amplitude** is 1.

sine rule The theorem that the lengths of the sides of any triangle are in the same proportion as the **sines** of the angles opposite the sides. If a, b, c in Figure S16 are the lengths of the sides opposite the angles A, B, C, then:

$$\frac{a}{\sin A} = \frac{b}{\sin B} = \frac{c}{\sin C}$$

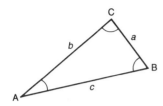
Figure S16

sinusoidal (a) [sy-nu-*soy*-dl] Shaped like the **sine curve**.
Example
The graphs of $2\sin x$, $\sin 3x$, $1 + \sin x$, $\cos x$ are all sinusoidal. Many wave motions in nature and science are sinusoidal.

skew (a) **1** (geometry)—Not in the same plane. If a pair of lines is skew, the lines are neither parallel nor intersecting, e.g. the diagonals drawn on the cube shown in Figure S17 form a skew pair.
2 (statistics)—Of a **distribution**, not symmetrical. Figures S18 and S19 illustrate skew distributions: Figure S18 is positively skewed, Figure S19 is negatively skewed.

Figure S17

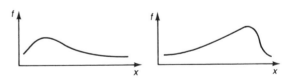
Figure S18 Figure S19

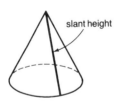
Figure S20

slant height The distance from vertex to base of a right circular **cone** (Figure S20) or of a right **pyramid** measured along the surface (Figure S21).

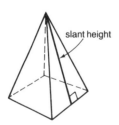
Figure S21

slide A term used in motion geometry to describe the effect of a **translation**.

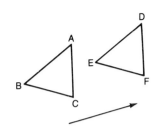

Figure S22

Example
Figure S22 represents the slide of a triangle from position ABC to position DEF. Note that after the slide each side remains parallel to its former direction.

slide rule   An instrument for carrying out multiplication, division and other mathematical operations. It consists basically of two **logarithmic scales**, which slide relative to each other. By this means, multiplication of two numbers is reduced to addition, and division is simplified to subtraction, as a consequence of the laws of logarithms. Slide rules were much used by scientists and engineers until the introduction of electronic calculators.

slope   Another word for **gradient**.
   The slope of a line on a **Cartesian** graph is the tangent of the angle it makes with the positive sense of the x-axis (Figure S23).
   The slope of a curve at a given point is the tangent of the angle made by the tangent to the curve at that point (Figure S24).

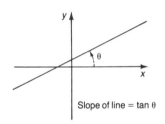

Figure S23

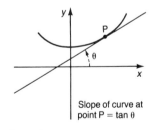

Figure S24

small circle   A circle drawn on a sphere but whose centre is not at the centre of the sphere. *Compare* **great circle**.
   Between any two points on the surface of a sphere, one great circle and an infinite number of small circles can be drawn. Any small circle path between the two points is longer than the great circle path, a fact that is taken into account when navigating on the surface of the earth.

solid   In geometry, a figure having three dimensions.
   Some of the solids described in this dictionary are **cube, cuboid, decahedron, tetrahedron,** other **polyhedra, cone, cylinder, prism, pyramid, ellipsoid, sphere**.

Sometimes, a solid is defined as a closed **surface** in space.

solid angle A surface consisting of **rays** starting from a common point (the vertex) and passing through a closed curve or plane figure. It is the three-dimensional equivalent of the ordinary two-dimensional **angle** composed of two rays.
Example
The curved surface of a cone forms a solid angle at the vertex of the cone, and in a similar way the four faces of a square pyramid (see Figure S25).

The unit for measuring solid angles is the **steradian**.

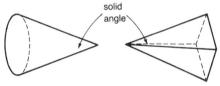

Figure S25

solid of revolution A solid formed by rotating a plane figure about a line in the plane of the figure. The line is called the axis of revolution.
Examples
a A **sphere** is formed when a circle is rotated about a line passing through the centre.
b A **cone** is formed by rotating a right-angled triangle about one of its shorter sides.
c A **cylinder** results from rotating a rectangle about one of its sides.

Any cross-section of a solid of revolution perpendicular to the axis of rotation is a circle.
See also **ellipsoid, paraboloid, spheroid, torus**.

solution The set of values that result in a true statement when replacing the unknowns in an **equation**.

Examples
a 3 is the only value for x that makes $2x + 7 = 13$ a true statement; so $x = 3$ is the solution to the equation.
b Either 1 or 2 may replace y in the equation $y^2 - 3y + 2 = 0$ to produce a true statement; so the solution set here is $y = 1, y = 2$.
Solution may also refer to the procedure used to **solve** an equation.
LATIN *solutio*: weakening

solve (v)
1 Of an equation, to find the value or values that **satisfy** the equation.
See **solution**.
2 Of a triangle, to find the sizes of all the angles and sides when information is available about only some of them.
See **cosine rule, sine rule**.
LATIN *solvere*: to loosen

space
An undefined concept in **Euclidean** geometry.

All geometric objects (such as points, lines, surfaces, solids) exist in space, and a three-dimensional object (a **solid**) is said to occupy a certain amount of space, called its **volume**.

One way of marking position and measuring distance in space is by **Cartesian coordinates** using three axes at right angles to each other.

In philosophy and in higher mathematics, the relation between space and time is explored, and systems of mathematics have been developed to describe **hyperspaces** having more than three dimensions.
LATIN *spatium*: space, room, extent

speed
The **rate** at which a moving object changes position with respect to time, e.g. a runner who covers 200 metres in 20 seconds has an average speed of $200/20 = 10$ metres per second.

sphere There are two definitions of a sphere:
1 The set of points in space that are all the same distance (r) from a fixed point called the centre. In this case, a sphere is a three-dimensional closed **surface**. Its area is given by the formula $A = 4\pi r^2$.
2 All the points enclosed by the surface defined above. In this case, a sphere is a **solid**. Its volume is given by the formula $V = \frac{4}{3}\pi r^3$.

Any plane **section** of a sphere is a circle. If the section passes through the centre of the sphere, the circle is a **great circle**; otherwise it is a **small circle**.
GREEK *sphaira*: a ball

Figure S26

spherical triangle A closed figure formed on the surface of a **sphere** from the **minor arcs** of three **great circles** (see Figure S26).

Unlike an ordinary plane triangle, the three angles of a spherical triangle can add up to more than two right angles. In fact, their sum can have any value from two right angles to six right angles.

spheroid A solid formed by rotating an **ellipse** about its major axis or minor axis. Its **cross-sections** are circles or ellipses.
Also, called **ellipsoid**.
See also **oblate**.
GREEK *sphaira*: a ball, *-o-eides*: like

spiral A plane curve traced out by a point winding around a fixed point according to some rule.
Example
In an **Archimedean spiral**, the distance of the moving point from the fixed point is proportional to the angle through which it has turned.

A **helix**, which is three-dimensional, is sometimes called a spiral.
LATIN *spira*: a coil

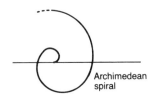

Figure S27

spread In statistics, an indefinite term describing the extent to which values in a **frequency distribution** are distributed about a central value such as the **mean** or **median**. *Compare* **dispersion, range**. *See* **interquartile range, standard deviation**, for more precise measures.

spreadsheet Paper ruled in columns and rows and prepared with headings and labels to help in the management of statistical data, e.g. the data may refer to cash flow in a business firm. Computers are often used for spreadsheet management.

square
1 A **rectangle** with equal sides, or a **rhombus** with equal angles. It follows from either definition that a square has four equal sides and four right angles, and that its diagonals are equal and bisect each other at right angles (Figure S28).
2 Of a number, the number multiplied once by itself, e.g. the square of 4 is $4 \times 4 = 16$. The square of n is written n^2 and read 'n squared' or 'the second power of n'.
3 A unit of area measurement formerly used by builders, equal to 100 square feet. It is now used by some builders for 10 square metres (10 m^2).

Figure S28

square (a)
1 Square **bracket**: [or]
2 Square measure: any system of units for measuring **area**, e.g. the **SI** unit of square measure is square metre (m^2).
3 Specifying the length of each side of a square, e.g. 'a room 3 metres square' refers to a square room that measures 3 metres by 3 metres. (Its area is 9 m^2.)
4 Square number: a whole number that is the square of a whole number, e.g. 1,4,9,16, ... are square numbers.

See also **figurate numbers**.
5 Square root: the number of which a given number is the square, e.g. 7 is the square root of 49, since $7 \times 7 = 49$. The square root of n is written $\sqrt{n}$ and read 'square root of n' or 'root n'.

square (v)
1 To square a circle: to construct a square equal in area to a given circle, using only straight-edge and compasses. This is a very ancient problem, but it was not until the late 19th century that it was proved to be impossible. Its impossibility arises from the fact that π is a **transcendental** number, and therefore a length equal to $\sqrt{\pi}$ cannot be constructed (see Figure S29).

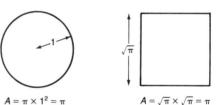

Figure S29

2 To square a number: multiply a number by itself once, e.g. to square 5 is to calculate $5 \times 5 = 25$.

standard deviation A measure of **dispersion** of a **frequency distribution**.
It is calculated from the individual deviations from the **arithmetic mean** by squaring these deviations, finding the average of the squares, and taking the square root of the average. The standard deviation is also known, therefore, as the root mean square deviation.
In a **normal distribution**, approximately 68% of the total distribution falls within one standard deviation either side of the mean, and approximately 95% falls within two standard

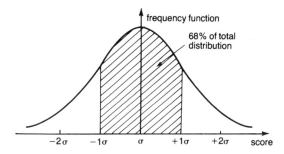

Figure S30

deviations either side of the mean.
Figure S30 shows the **normal curve**, with the score axis marked in numbers of standard deviations (σ).

standard error A way of estimating how close the mean of a **sample** is to the mean of the whole population. It is based on a knowledge of the **standard deviation** (σ) of the population and the size (N) of the sample:

Standard error = $\sigma/\sqrt{N}$

Example
Suppose that on a nation-wide mathematics test of 16-year-old students, the scores showed a standard deviation of 15 and that the mean score for one school sample of 100 students was 60. The standard error for this mean is $15/\sqrt{100} = 1.5$. Referring to the entry under standard deviation, it can be seen that there is a 68% chance that the population mean lies within one standard error of 60 (i.e. between 58.5 and 61.5), and that there is a 95% chance that it lies within two standard errors (i.e. between 57 and 63). This calculation gives a measure of confidence of how far the sample mean can be accepted as representative of the population mean.

standard form Another name for **scientific notation**, e.g. written in standard form, the

standard score 218

number 123.5 is 1.235×10^2 and the number 0.576 is 5.76×10^{-1}.

standard score A score in a **frequency distribution** converted to show the number of **standard deviations** it lies from the mean of the distribution.
Example
The following table shows some standard scores for a distribution whose mean is 60 and standard deviation is 10:

actual score:	60	70	80	50	45
standard score:	0	1	2	−1	−1.5

standard unit A base unit in any system of units of measurement. See especially **Système international** (SI), which has the following standard (or base) units: metre, kilogram, second, ampere, kelvin, candela and mole.

statics [*stat*-iks] The study of forces acting on a body in a state of equilibrium or rest.
Compare **dynamics**.
GREEK *statos*: standing

stationary point A point on the graph of a function at which the **tangent** to the graph is parallel to the x-axis. At a stationary point, the gradient is zero.

A stationary point may be at a **maximum** value of the function (see Figure S31), or at a

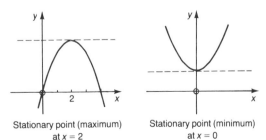

Stationary point (maximum) at $x = 2$ Stationary point (minimum) at $x = 0$

Figure S31   **Figure S32**

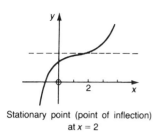

Stationary point (point of inflection) at $x = 2$

Figure S33

minimum value (Figure S32), or it may be a **point of inflection** (Figure S33).

statistic [sta-*tis*-tik]   In **statistics**, any characteristic of a **sample** calculated from observations and used to estimate the corresponding characteristic of the **population** from which the sample has been drawn. Examples of a sample statistic are **mean, median, standard deviation**. Corresponding to each sample statistic, there is a population **parameter**.

statistics [sta-*tis*-tiks]   A branch of mathematics concerned with the systematic collection and arrangement of large numbers of observations and quantities of numerical observations, and with ways of drawing useful conclusions from such data.
  Descriptive statistics includes ways of managing a large body of data using calculated measures such as **mean, percentiles, standard deviation** to summarise tendencies, and graphical methods like **histograms** to present characteristics visually.
  Inferential statistics is concerned with making statements about large **populations** based on evidence drawn from **samples**.
  Central to the study of statistics are the mathematics of **correlation** and of **probability** and **chance**.

stem and leaf plot   In statistics, a way of organising and presenting a collection of numbers.
Example
The set of numbers on the left (below) are rewritten at the right, each tens digit becoming a 'stem' and the units digits becoming the 'leaves'. The leaves have been entered in numerical order to produce an ordered stem and leaf plot:

				Stem	Leaf
57	63	54	69	4	6 8 8 8 9
49	48	61	70	5	2 4 7
48	46	52	60	6	0 1 3 3 6 9
63	74	48	66	7	0 4

steradian [ste-*ray*-di-an]   A unit for measuring a **solid angle**. Symbol sr. One steradian is the measure of a solid angle that has its vertex at the centre of a unit sphere and cuts off unit area at the surface of the sphere. Because the total surface area of a unit sphere is 4π, the largest possible solid angle is an angle of 4π steradians.
GREEK   *stereos*: solid, + radian

stochastic (a) [sto-*kas*-tik]   Depending on chance. A **random variable** is sometimes called a stochastic variable.
GREEK   *stochos*: aim

straight (a)   **1**   Of a **line**, having only one direction. In geometry, a line is sometimes referred to as a straight line to distinguish it from curves. **2**   A straight angle is an angle of 180 degrees or two right angles. **3**   A straight edge is an instrument used to draw lines — like a ruler without graduations.

submultiple   The same as **factor**, e.g. 2,3,4 and 6 are submultiples of 12.
LATIN   *sub–*: under

subset   A **set** whose members are all members of another set.
Example
If A = {2,4,6,8,10} and B = {2,4,6,}, then B is a subset of A. This is written B ⊂ A, and B is said to be a 'proper' subset of A because it is not equal to A.
  Any set is said to be a subset of itself (but not a proper subset).
  The **empty set** is a subset of every set.

substitute (v) In algebra, to replace part of an expression by another expression or number, usually with intent to **simplify** or **evaluate**.
Examples
a Substituting $y = x + 1$ in the equation $(x + 1)^2 + 2(x + 1) - 3 = 0$ gives $y^2 + 2y - 3 = 0$, which is a simpler equation to solve.
b If $a = b^2 + 1$, various values of a are found by substituting various values for b: if $b = 3$, $a = 10$; etc.
LATIN *sub*–: under, *statuere*: to set up

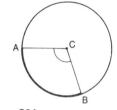

Figure S34

subtend (v) To stretch under, e.g. in Figure S34, arc AB subtends angle ACB at the centre of the circle.

The diameter of the moon subtends an angle of about half a degree at the eye of an observer on earth.
LATIN *sub*–: under, *tendere*: to stretch

subtraction; subtract (v) Subtraction is one of the four fundamental operations of arithmetic. (The others are **addition**, **multiplication** and **division**.) At the simplest level, to subtract means to take away, as in $9 - 4 = 5$, where the minus sign is used for the operation. At a less simple level, a number can be subtracted from a smaller number to give a negative number, as in $5 - 8 = -3$, and a negative number can be subtracted as in $7 - (-3) = 10$.

Subtraction is the **inverse** of **addition**: $7 - 3$ is the solution to the equation $3 + x = 7$.

The rules for subtraction of **signed numbers** are:
- $a - (-b) = a + b$
- $(-a) - b = -(a + b)$
- $(-a) - (-b) = b - a$

LATIN *sub*–: under, *trahere*: to drag away

subtrahend A number subtracted, e.g. in the calculation $12 - 3 = 9$, 3 is the subtrahend.

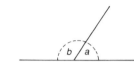
Figure S35

Figure S36

Figure S37

sum 1 The result of adding numbers or quantities together.
Examples
a The sum of 5 and 7 is 12.
b The sum of $8.50 and $1.50 is $10.00.
c The sum of a, b and c is $a + b + c$.
The Greek letter, capital sigma (Σ), is used for the sum of terms in a **series**.
2 Another name for the **union** of two sets.
LATIN *summa*: top, summit

supplementary angles A pair of angles whose sum is two right angles (180°). Each angle of the pair is the supplement of the other.
Example
Figure S35 shows a straight line, Figure S36 a parallelogram, and Figure S37 a cyclic quadrilateral. In each case, the angles marked a and b are supplementary: $a + b = 180°$.
Compare **complementary**.
LATIN *supplere*: to fill up, complete

surd An **irrational** number that is also the **root** of a whole number or fraction, e.g. $\sqrt{3}, \sqrt[3]{5}$. An expression containing one or more such numbers, e.g. $2 + \sqrt{5}, \sqrt{2} + \sqrt[3]{7}$.
LATIN *surdus*: deaf

surface A continuous two-dimensional portion of space.
The boundary of a solid forms a closed surface. It may consist of planes (e.g. a **cube**), it may be curved (e.g. a **sphere**), it may be partly plane and partly curved (e.g. a **cylinder** or a **cone**).
See also **area**.
LATIN *super*: over, *facius*: face

survey A general view or description.
Examples
Traffic surveys, health surveys and opinion surveys are examples of statistical surveys.

a   A traffic survey is a count of the number and type of vehicles using the road.
b   A health survey is a collection of data relating to the health of a population.
c   An opinion survey (or poll) is an analysis of public opinion on some subject.

Another type of survey is a land survey. This is carried out by measuring distances and slopes of a piece of land. A person trained to do this is called a surveyor.

LATIN   *super*: over, *videre*: to see

symbol [*sim*–bl]   A mark or sign standing for some thing or some idea. Many symbols are used in mathematics. They are used for
- **numbers** e.g. 5, 3.2, π
- **variables** e.g. x, θ
- **operations** e.g. $+, -, \times, \div, \sqrt{\ }$
- **relations** e.g. $=, >, \cup$
- **functions** e.g. log, sin
- **points, lines, angles** e.g. P, AB, $\angle$PON etc.

GREEK   *symbolon*: a token

symmetric (a) [sim–*met*–rik]   Describing a relation that remains true when the related quantities change place.
Example
'Equals' is a symmetric relation: if a equals b then also b equals a.
   'Is less than' is not a symmetric relation: if a is less than b, then b is not less than a.

GREEK   *sym*–: with, *metron*: measure

symmetrical (a) [sim–*met*–rik–l]   (sometimes, symmetric).
1   (geometry) — Balanced about a line or point.
Examples
a $\sqcap\!\!\sqcup$ is symmetrical about the dotted line.

b Z is symmetrical about the point marked.

See also **axis, line of symmetry, rotational symmetry**
2 (algebra) — Of an expression, the same in value when two variables are interchanged, e.g. $2x^2 + 3xy + 2y^2$ is **symmetrical**; $2x + y$ is not symmetrical.

symmetry [*sim*–met–ri] In geometry, the property of being balanced about a point, line or plane.
See **axis, line of symmetry, rotational symmetry, symmetrical**.

syntax [*sin*–taks] The study of the rules governing the order of symbols used in any language; in particular, in the languages of mathematics, logic and computers.
GREEK *syntaxis*: arrangement

Système International d'Unités
Abbreviation **SI**. A world-wide system of units of measurement. It is designed for the measurement of all physical quantities and, as well as being used in science and technology, it is also in general use. There are seven base units from which all other units are derived:

metre (m) for	length
kilogram (kg) for	mass
second (s) for	time
ampere (A) for	electric current
kelvin (K) for	temperature
candela (cd) for	intensity of light
mole (mol) for	amount of substance

The base units and the many derived units (e.g. for area, volume, velocity) may be used alone or with prefixes that form multiples and submultiples. A full list of these prefixes is contained in the Appendix.
FRENCH International System of Units

T

tally A mark made to keep count of a number of objects or events. Tally marks are often grouped together in fives to help in counting the total. For example, the tally sheet for counting the results of tossing two coins may look like this:

Result	Tally	Frequency
2 heads	⊬⊦⊦ I	6
head and tail	⊬⊦⊦ ⊬⊦⊦ II	12
2 tails	⊬⊦⊦	5

LATIN *talea*: a rod or stick

tan Abbreviation and symbol for **tangent**, used in trigonometry and coordinate geometry.

tangent [*tan*-jent]
  1 (plane geometry) — The tangent to a curve at a point on the curve is the (straight) line having the same **direction** as the curve at that point.

At a point where the curve is **concave** or **convex**, the tangent touches the curve but does not intersect it. At a **point of inflection**, the tangent intersects the curve. Figure T1 illustrates these three possibilities.

Figure T1

An important case is the tangent to a circle: the tangent at any point on the circumference of a circle is at right angles to the radius drawn to that point (see Figure T2).

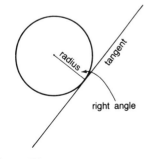

Figure T2

tangent

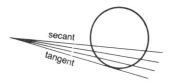

Figure T3

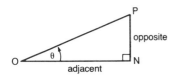

Figure T4

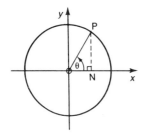

Figure T5

One way of regarding the tangent to a circle is to see it as the limiting position of a series of **secants** drawn from a point outside the circle. Each secant cuts the circle at two points: the tangent touches the circle at one (see Figure T3).

The point shared by a curve and its tangent is called the point of contact.

2 (solid geometry) — In the same way that a tangent line can be drawn to a curve at a point, a tangent plane can be drawn to a curved surface at a point. The tangent plane to a sphere is at right angles to the radius drawn to the point of contact.

3 (trigonometry) — One of the **trigonometric ratios** associated with an angle: in Figure T4, $\tan \theta = PN/ON$.

Compare **cosine, sine.**

4 (coordinate geometry) — One of the **circular functions**: The tangent of the angle θ in the circle shown in Figure T5 is the ratio of the projection of OP on the y-axis ($=y$) to the projection of OP on the x-axis ($=x$); that is $\tan \theta = y/x$. This value is positive when P is in the first or third quadrant, and negative when P is in the second or fourth quadrant. As P moves around the circle in either direction, θ takes on values that lie between $-\infty$ and $+\infty$, and $\tan \theta$ also varies between $-\infty$ and $+\infty$, as suggested by the graph in Figure T6.

LATIN *tangens*: touching

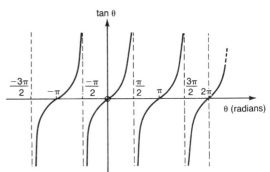

Figure T6

tangential (a) [tan-*jen*-shul]   Relating to a **tangent**, or in the direction of a tangent, e.g. the line in Figure T7 is tangential to the curve.

Figure T7

tangram   An ancient Chinese puzzle consisting of a square card or tile cut into seven pieces, which are then to be rearranged in a variety of shapes. Figure T8 shows the seven pieces forming the original square and then Figure T9 shows a rearrangement in the form of a man running.

Figure T8

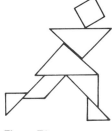
Figure T9

tera-   A prefix meaning one million million times. Symbol T.
Example
   1 terametre = 1 000 000 000 000 metres
      1 Tm = 10^{12} m
GREEK   *teras:* monster

term

1   An element of a **sequence** or of a **series**, e.g. each of the numbers separated by commas in the following sequence is a term: 1,3,5,7, ...

2   Part of an algebraic expression, especially a part separated from the rest by plus or minus signs, e.g. the expression $2x^2 + x - 10$ has the three terms: $2x^2$, x, and -10.

In the expression $(2x + 5)(x - 2)$ each bracket considered separately contains two terms; but, considered as a whole, the expression consists of only one term.

To **factorise** may be thought of as the process

of reducing a many-termed expression to a single term, as in:

$$2x^2 + x - 10 = (2x + 5)(x - 2)$$
(three terms) = (one term)

LATIN *terminus*: a boundary

tessellation An arrangement of plane figures, usually of the same shape and size, to cover a surface without gaps or overlapping. The polygons shown in Figure T10 are examples, but not all polygons can tessellate a plane, e.g. octagons cannot.

LATIN *tessella*: a small cube

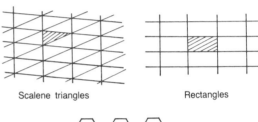

Scalene triangles Rectangles

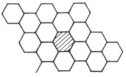

Figure T10 Hexagons

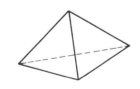

Figure T11

tetrahedron A solid with four plane faces. The faces are triangles. For a **regular** tetrahedron, the faces are **equilateral** triangles (see Figure T11)

GREEK *tetra*: four, *hedra*: base

theodolite [the-*od*-o-lyt] A surveyor's instrument for measuring angles in a horizontal or vertical plane.

theorem A proposition (statement) that can be proved by arguing logically from **axioms** or other statements assumed or known to be true. *See* **proof**.

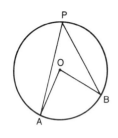

∠AOB = 2∠APB

Figure T12

Example
Many mathematics textbooks include a proof that the angle at the centre of a circle is double the angle at the circumference standing on the same arc (see Figure T12).

Some theorems in mathematics have had to wait a long time for proof. *See*, for example, the **four-colour theorem**.

three-dimensional (a) Referring to ordinary space, which is called three-dimensional because to locate a point in space requires three **coordinates**.

A common system of coordinates is based on three axes at right angles to each other. In Figure T13, point P has the coordinates (2,4,5).

Solid figures, such as a cube and a sphere, are three-dimensional.

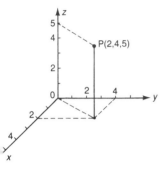

Figure T13

time interval The lapse of time between two moments.
Example
If a heart beats 70 times in a minute, the average time interval between one heart beat and the next is 1/70 of a minute.

time line A **number line** on which the numbers represent years.

tonne Abbreviation: t A non-SI unit of **mass**.
 1 tonne = 1000 kilograms = 1 megagram (Mg)

topology [to-*pol*-o-jee] The study of those properties of geometric figures that remain unchanged when the figures are bent, stretched or twisted without tearing. Topology is concerned with relative positions, but not with distances or angles.

A cube, a tetrahedron and a sphere are topologically equivalent: they each divide space into two regions—inside and outside.

Transport networks in large cities are often

advertised in the form of topologically equivalent diagrams, in which the order of stations and the connections between lines are preserved, while directions are not necessarily accurate and distances are not to scale.
GREEK *topos*: place, *logos*: a study

torus The solid formed when a closed curve (usually a circle) is rotated around an axis that is in the same plane as the circle but is outside the curve.

It is also called an anchor ring. A quoit and a doughnut have the shape of a torus.
LATIN *torus*: a bulge

towers of Hanoi An old puzzle consisting of three posts fixed to a base, together with a set of rings of different diameters. To start with, the rings are placed on one of the poles in decreasing order of size. The aim of the puzzle is to move the rings, one at a time, so that they end up in the same order on another post; but at no time must a ring be placed above one that is smaller.

With n rings, the minimum number of moves to achieve this is $2^n - 1$.
Also known as towers of Atlantis.

Figure T14

trajectory [tra-*jek*-to-ree] **1** The path of a projectile after it is launched, e.g. the path of a ball after it is thrown; the path of a bullet shot from a gun. The path of a planet around the sun is sometimes referred to as a trajectory, though it is more usually known as an orbit. **2** A curve that cuts every member of a **family** of curves at right angles.
LATIN *trans*: across, *jacere*: to throw

transcendental (a) [tran-sen-*den*-tl] Not algebraic.
1 A transcendental number is one that cannot

be the solution of an **algebraic equation** with rational coefficients. It is one kind of **irrational number**. Transcendental numbers especially mentioned in this dictionary are π (**pi**) and **e**, but others occur often in mathematics.
2 Transcendental functions are not algebraic, e.g. $\sin x$, $\log x$, 2^x are transcendental functions of x.
LATIN *transcendere*: to step over

transform (v) To change the arrangement of a formula or expression without changing its value or meaning. An example of transforming a formula is **changing the subject**, e.g. $A = \pi r^2$ becomes $r = \sqrt{A/\pi}$.
LATIN *trans*: across, *forma*: shape

transformation A **mapping** of one geometrical figure to another according to some rule. For examples of transformations, see **dilation, reflection, rotation, translation**.

translate (v) To slide a geometric figure from one position to another without rotating it or changing its shape or size.

translation A **transformation** of a geometric figure in which every point is moved the same distance in the same direction.

Example
In Figure T15, Figure A is a translation of Figure B, and B is a translation of A.
  In coordinate geometry, a translation of axes is sometimes carried out to simplify the equation of a function.

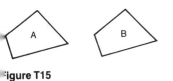
Figure T15

Example
In the translation illustrated in Figure T16, the axes are moved 3 units in the x-direction, thus changing the equation from $y = (x - 3)^2$ to $y = x^2$

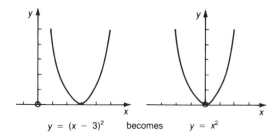

Figure T16

transpose   **1** To move a term from one side of an **equation** to the other, with change of sign.
Example
The equation $2a = 3b - 1$ becomes $2a - 3b = -1$ when the term $3b$ is transposed from right to left. The two forms are equivalent.
2 In matrix algebra, to interchange rows and columns in a matrix.
Example

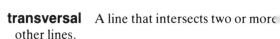

FRENCH   *transposer*: to put across

transversal A line that intersects two or more other lines.
Example
In Figure T17, PQ is a transversal for the other three lines shown.

If **parallel** lines are intersected by a transversal, the corresponding angles at the points of intersection are equal.
LATIN   *transversus*: lying across

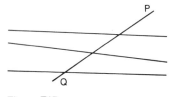

Figure T17

trapezoid A **quadrilateral** having one pair of opposite sides parallel and unequal.
LATIN   *trapezium*: a small table

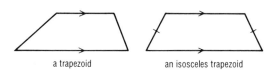

a trapezoid   an isosceles trapezoid

Figure T18

tree diagram

Figure T19

trapezium A **quadrilateral** with neither pair of opposite sides parallel.

travel graph A graphical representation of a journey, produced by plotting the traveller's **displacement** against the time of travel.
Example
Figure T20 illustrates the following journey. A motorist travels at constant speed for 1 hour reaching a distance of 50 km from the starting point, then rests for ½ hour before resuming the journey at increased speed to arrive at a point 100 km from home after a total of 2 hours. This is followed immediately by a slower return journey, with the graph showing the motorist 50 km from home after 4 hours from the start.

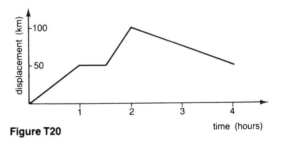

Figure T20

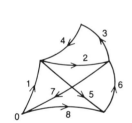

Figure T21

traversable (a) In **topology**, describing a **network** that can be traced without lifting the pencil or retracing an arc.
Example
Figure T21 shows a traversable network: it can be traced by starting at 0 and moving along the arcs in the order as numbered.
Figure T22 is not traversable.
FRENCH *travers*: across

Figure T22

tree In **topology**, a tree-shaped diagram, as in Figure T23.

tree diagram A branching graph without loops, representing the possible outcomes in a probability experiment.

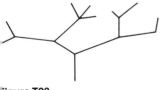

Figure T23

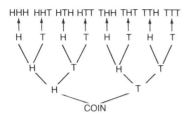

Figure T24

Example
The tree diagram in Figure T24 shows the eight possible outcomes (the **sample space**) of tossing a coin three times. From the diagram, it can be seen that each of the outcomes in this case is equally likely, so that the probability of any one of the results is 1/8.

trial   In statistics, a single **event** or observation, e.g. if an experiment consists of throwing a die many times, then one throw is a trial.

triangle   A closed figure with three straight sides. It follows that a triangle is a plane figure. The three angles of any triangle add up to two right angles (180°, π radians).

The area of any triangle is found by halving the product of the length of one of its sides by the length of the **altitude** drawn to that side.

Triangles are classified according to their sides, as illustrated in Figure T25, or in terms of their angles as in Figure T26.

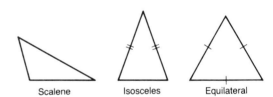

Figure T25

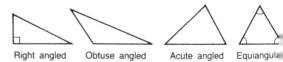

Figure T26

Triangles are of great theoretical importance in mathematics, and they have important practical applications in surveying and construction engineering.
LATIN   *triangulus*: three-cornered

triangular number A whole number that can be represented by the number of dots needed for a triangular array, the dots being evenly spaced.
Example
The first three triangular numbers are 1,3 and 6 as shown in Figure T27.

Figure T27

The nth triangular number is the sum of the first n whole numbers:

$$\begin{aligned} 1 &= 1 \\ 1+2 &= 3 \\ 1+2+3 &= 6 \\ 1+2+3+4 &= 10 \\ 1+2+3+4+5 &= 15 \\ \text{etc.} \end{aligned}$$

See also **figurate numbers, Pascal's triangle**.

triangulation A way of surveying a piece of land. A base line is measured and landmarks are sighted from points on this line. The angles between the base line and the lines of sight are measured. Distances between the landmarks can then be calculated using the formulae of **trigonometry**.

trigonometric functions The following functions: **sine** (abbreviated sin), **cosine** (cos), **tangent** (tan), and their inverses, **cosecant** (cosec), **secant** (sec), **cotangent** (cot). Also called **circular functions**.

Trigonometric functions are important in the mathematical description of many periodic phenomena in science, engineering, etc.
See separate definitions for each of the above.

trigonometric ratios The ratios that define the **trigonometric functions** as properties of an

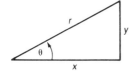

Figure T28

angle in a right-angled triangle (see Figure T28), as follows:

$$\sin \theta = y/r \quad \operatorname{cosec} \theta = r/y$$
$$\cos \theta = x/r \quad \sec \theta = r/x$$
$$\tan \theta = y/x \quad \cot \theta = x/y$$

Párticular values for the trigonometric ratios of 30°, 60°, 45° can be seen from the triangles in Figure T29:

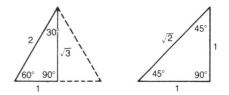

Figure T29

	0°	30°	45°	60°	90°
sine	0	1/2	$1/\sqrt{2}$	$\sqrt{3}/2$	1
cosine	1	$\sqrt{3}/2$	$1/\sqrt{2}$	1/2	0
tangent	0	$1/\sqrt{3}$	1	$\sqrt{3}$	∞

trigonometry The branch of mathematics concerned with the properties of triangles and calculations based on these.

Trigonometry has many direct practical applications, including surveying and navigation, and, through the **trigonometric functions**, is the basis for much higher mathematics. Spherical trigonometry is a development concerned with measurements on a sphere.

See **spherical triangle**.

GREEK   *tri*: three, *gonia*: angle, *metron*: measure

trillion   In North American usage, 1 trillion = thousand 'billion' = 1 thousand thousand million = 10^{12}.

In British usage, 1 trillion = 1 million 'billion' 1 million million million = 10^{18}.

trinomial An algebraic expression with three **terms**, e.g. $2x^2 + 3x - 1$; $a - b + c$.
LATIN *tri*: three, *nomen*: a name

trisect (v) In geometry, to divide into three equal parts such figures as a line segment or an angle.
LATIN *tri*: three, *secare*: to cut

truncated (a) Of a geometric solid, having the **apex** removed by slicing through the solid with a plane cut (see Figure T30). If the cutting plane is parallel to the base, the result is a **frustum**.
LATIN *truncare*: to cut off

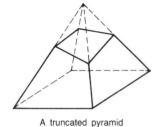

A truncated pyramid

Figure T30

truth table In **logic**, a table setting out the various true and false possibilities of combinations of **propositions** each of which may be true or false.
Example
In the following table, T and F stand for 'true' and 'false' respectively, showing the truth values for the **conjunction**, p∧q, which correspond to the separate truth values of p and q.

p	q	p∧q
T	T	T
T	F	F
F	T	F
F	F	F

turning point A point on a graph where the **gradient** is zero and lying between points for which the gradients are of opposite signs.
Example
In Figure T31 A and B are turning points. The graphed function has either a **maximum** or a **minimum** value at the turning point. A turning point is an example of a **stationary point**; the other kind of stationary point is a **point of inflection**.

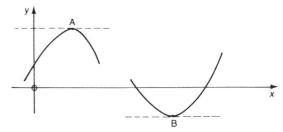

Figure T31

two-dimensional (a)   Referring to the geometrical condition that requires two **coordinates** (or **dimensions**) to specify the position of a point. The position of a point that is restricted to a **plane** can be stated with two numbers, as, for example, with **Cartesian** coordinates or with **polar coordinates**.

Plane figures are two-dimensional, e.g. polygons, circle, ellipse. Contrast these with **line segments**, which are one-dimensional and with **solids**, which are three-dimensional.

U

union In set theory, the set of elements made up from the elements of a pair of sets.
Example
The union of set A {1,2,4,8} and set B {0,1,2,3} is the set {0,1,2,3,4,8}. This is written A ∪ B and read 'A cup B' or 'A union B'.
Also called **sum**.
LATIN *unus*: one

unique (a) [yoo-*neek*] Being one of a kind, e.g. the equation $2x + 3 = 11$ has a unique solution, $x = 4$. There are no other solutions.
 The square root of 16 does not have a unique value: there are two values, $+4$ and -4.
LATIN *unus*: one

unit 1 A single thing. The first place in a whole number as written in a system that uses **place value** tells the number of units, e.g. 139 in the **decimal** system has nine units, three tens and one hundred; 11 in the **binary** system has one unit and one two.
2 A defined amount of a quantity, serving as a basis for measuring other amounts of the same quantity, e.g. the metre is defined as a unit of length, and then the length of a race track may be measured as 100 metres.
See **Système International d'Unités**.
LATIN *unus*: one

unitary method [*yoo*-ni-tair-ee] A method of solving **proportion** problems of the following kind:
Example
If 8 items cost $24, what is the cost of 5 items?
 The method consists of first finding the cost of

1 item and then multiplying by the second number of items:
$$\$24 \div 8 = \$3, \$3 \times 5 = \$15$$
LATIN *unus*: one

unit circle A circle with a radius of 1.

unity The number one.

universal set The set that defines all the members relevant to a particular investigation, e.g. if there is a discussion about the membership of committees in a club, then the universal set is the set of all club members. All the sets under discussion are **subsets** of the universal set. Any set and its **complement** make up the universal set.

The symbol for universal set is U.

A **Venn diagram** shows the relationship between the universal set and other sets.

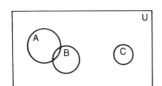

Figure U1

unknown In an **equation**, a variable whose value (or values) is yet to be found to make the equation true.
Examples
a $2x - 3 = 0$ is an equation in one unknown (x).
b $2x + 5y - 7 = 0$ is an equation in two unknowns (x, y).

V

valid (a) [*val*-id]   In **logic**, describing a sound argument or **inference**.
Being true and being valid should not be confused. It is possible for a valid argument to lead from a false **premise** to an untrue conclusion, just as it is possible for an argument that is not valid to lead to a true conclusion.
LATIN   *validus*: strong

value   1   A number put in place of a variable in a formula or function, e.g. a sequence of numbers can be formed from $2n - 1$ by replacing n by the positive integers: giving n the values, $1,2,3,\ldots$ produces the sequence $1,3,5,\ldots$
2   A number that can replace an unknown in an equation to make the equation true, e.g. $3x - 2 = 13$ has the solution $x = 5$; this is the value of x that makes the equation true.
LATIN   *valere*: to be worth

vanish (v)   To be zero, or to approach zero, e.g. $2n - 1$ vanishes (equals zero) for $n = \frac{1}{2}$; $1/x$ vanishes (approaches zero) as x becomes very large.

variable   A quantity, represented by a symbol, that can take on any one of a set of values.
Examples
a   In an **equation**: in $3a + 2 = 5$, a is a variable. If its value is 1, the statement is true. If a has any other value, the statement is false.
b   In a **formula**: in the formula $c = 2\pi r$, c and r are variables, whereas π is a **constant**. The formula enables one to calculate the circumferences (c centimetres) of circles having different radii (r centimetres). The

formula is true for all positive values of r.

c In the study of **functions**: $y = 2x^2$ represents the functional relation between an **independent** variable, x, and its **dependent** variable, y. x may have any value and the values of y depend on the values chosen for x.

LATIN *variare*: to change

variation A relation between two or more variables:
- Direct variation (*see* **direct proportion**) is represented by $y \propto x$ or $y = kx$.
- Inverse variation (*see* **inverse proportion**) is represented by $y \propto 1/x$ or $y = k/x$.
- **Joint variation** is represented by $z \propto xy$ or $z = kxy$.
- Variation as the inverse square is represented by $y = k/x^2$.

In these cases, x, y, z are variables and k is a constant.

LATIN *variare*: to change

variance A measure of **dispersion** of a **frequency distribution**. It is calculated from the individual deviations away from the **arithmetic mean** by squaring these and finding the average of the squares. The variance is thus the mean square deviation and is the square of the **standard deviation**.

The usual symbol for variance is σ^2 for a **population** and s^2 for a **sample**.

vector A quantity represented by a **line segment** having both **magnitude** and **direction**.
Example
Displacement, velocity, acceleration, force are examples of vector quantities of importance in applied mathematics. They contrast with **scalar** quantities, which have only magnitude and no direction property.

There are special rules for carrying out operations on vectors.

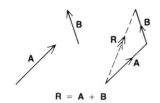

Figure V1

Example
Figure V1 shows how to find the result of adding vector **A** to vector **B** to get the sum (or resultant) **R**.

Some symbols for vectors: $\overrightarrow{AB}$, $\vec{F}$, $\overline{F}$, **F** — the last is the symbol we have used in this book.
See also **resolution, resultant.**
LATIN *vector*: carrier

Venn diagram A diagram showing the relation between sets drawn as circles inside a boundary representing the **universal set**.
Example
In Figure V2, U represents all people; M represents all male mathematicians; F represents all female mathematicians; S represents all scientists.

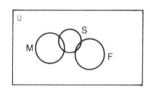

Figure V2

Areas have no significance in a Venn diagram, so this diagram says nothing about the numbers of mathematicians or scientists. It illustrates relations such as that some (but not all) scientists are mathematicians, that some people are neither mathematicians nor scientists, and (of course) that no male mathematicians are female.
See also **Euler diagram**.
Named after John Venn, England, 1834–1923.

vertex *pl.* **vertices** [*ver*-ti-seez] The point of intersection of the sides of a **polygon** or the faces of a solid, e.g. a triangle has three vertices and a cube has eight (Figure V3).

Figure V3

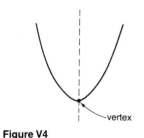

Figure V4

It is also the name given to the point of intersection of a **parabola** with its axis.
LATIN *vertex*: highest point

vertical (a) Upright; at right angles to the **horizontal**. A plumb-line is used to determine the vertical direction near the earth's surface.

The term is also applied to the *y*-axis of a **Cartesian** graph.

Vertical should not be confused with **perpendicular**. A perpendicular line is at right angles to another line or a plane and is not necessarily vertical.

LATIN *vertex*: highest point

vertically opposite (a) Describing the pairs of equal angles formed when two lines intersect.
Example
In Figure V5, *a* and *c* are a pair of vertically opposite angles; *b* and *d* are another pair; $a = c$; $b = d$.

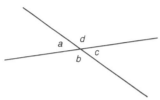

Figure V5

vinculum A stroke placed above two or more terms in an expression, used sometimes for the same purpose as **brackets**, e.g. $a - \overline{b + c}$ is the same as $a - (b + c)$.

LATIN *vinculum*: a bond

volume Amount of space occupied by a solid. It is a three-dimensional quantity (*compare* **area**, which is two-dimensional, and **length**, which is one-dimensional).

The unit of volume in most systems of measurement is the volume of a cube of length 1 unit, e.g. the fundamental **SI** unit of volume is 1 cubic metre (1 m^3).

Some useful volume formulae:
- cube $\quad V = a^3$
- rectangular prism (box) $\quad V = abc$
- sphere $\quad V = \frac{4}{3}\pi r^3$
- cylinder $\quad V = Ah$
- cone $\quad V = \frac{1}{3}Ah$

LATIN *volumen*: a roll

vulgar fraction The same as **common fraction**, that is, a fraction expressed as a ratio. Not a decimal fraction, e.g. 11/4, 7/2.

LATIN *vulgaris*: of the common people

W

weight The gravitational force by which an object is attracted to the earth. Different objects have weights in proportion to their masses if all are at the same distance from the earth's centre.

The word 'weight' is often used unscientifically instead of **'mass'**.

weighted mean or **weighted average** The **mean** (average) of a set of numbers after allowing for the numbers to have different effects. It is calculated by multiplying each by another number (called its weight), which is proportional to the desired effect, and dividing the sum of the weighted numbers by the sum of the weights.

Example

The annual salaries listed for the twenty workers in a small business are (in thousands of $) 40,35,25,20. The mean of these is 30; but a more useful weighted mean is obtained by first multiplying each salary by the number of workers drawing that salary:

salaries	40	35	25	20
workers	1	6	9	4
product	40	210	225	80

$$\text{weighted mean} = \frac{40+210+225+80}{1+6+9+4}$$
$$= \frac{555}{20}$$
$$= 27.75$$

whole number A **natural number** or counting number; a positive **integer**, e.g. 1,2,3,4,... Sometimes zero (0) is included. Sometimes negative integers (−1,−2,−3,...) are also included.

X

X-axis, Y-axis, Z-axis Names given to the axes in a **Cartesian** coordinate system.

When only two dimensions are needed, values of the **independent** variable are plotted in the direction of the X-axis (the left–right direction) and values of the **dependent** variable in the direction of the Y-axis (at right angles to the X-axis) (Figure X1).

For three dimensions, the Z-axis is constructed at right angles to the plane of the X- and Y-axes (Figure X2).

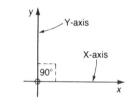

Figure X1

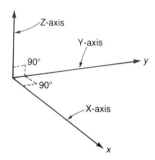

Figure X2

Z

Zeno's paradoxes Zeno of Elea, who lived in the 5th century BC, proposed a number of **paradoxes** that have been of interest to mathematicians because they contribute to an understanding of some fundamental concepts such as space, time, motion, limit, continuity, infinity. For a description of one of these paradoxes, see **Achilles and the tortoise**.

zero The number denoted by 0, also called nought. It stands for the absence of any **magnitude**.

It is one of the **Arabic numerals**, and is used to mark a place not occupied by any of the other numerals, e.g. when writing the number two hundred and seven, zero is used to indicate that there are no tens: 207.

Zero is the **cardinal** number of the **empty** set.

When zero is added to or subtracted from any number, that number is unchanged: $n + 0 = n$; $n - 0 = n$. When a number is multiplied by zero, the result is zero: $n \times 0 = 0$. Division by zero is not allowed.

Zero on the **number line** separates negative numbers from positive numbers. Zero itself is neither positive nor negative:

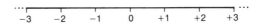

Through French (*zéro*) and Italian (*zerifo*) from Arabic *sifr*: cipher

Appendix 1 Table of SI prefixes
(Used with metric measures)

Prefix	Symbol	Value	
exa-	E	10^{18}	1 000 000 000 000 000 000
peta-	P	10^{15}	1 000 000 000 000 000
tera-	T	10^{12}	1 000 000 000 000
giga-	G	10^{9}	1 000 000 000
mega-	M	10^{6}	1 000 000
kilo-	k	10^{3}	1 000
hecto-	h	10^{2}	100
deka-	da	10	10
deci-	d	10^{-1}	0.1
centi-	c	10^{-2}	0.01
milli-	m	10^{-3}	0.001
micro-	μ	10^{-6}	0.000 001
nano-	n	10^{-9}	0.000 000 001
pico-	p	10^{-12}	0.000 000 000 001
femto-	f	10^{-15}	0.000 000 000 000 001
atto-	a	10^{-18}	0.000 000 000 000 000 001

Examples
a   microsecond μ'' one millionth of a second
b   kilometre   km   one thousand metres
c   megatonne   Mt   one million tonnes

Note: centi- and deci- are used only with metre.
hecto- and deka- are seldom used at all.

Appendix 2 Table of other prefixes used in mathematics

Prefix	Original language	English meaning	Example
a-	Greek	not	asymmetric
ad-	Latin	at; near; to	adjacent
bi-	Latin	twice; two	bisect
circum-	Latin	around	circumcircle
co-	Latin	with; share	coplanar
con-	Latin	with; share	converge
deca-	Greek	ten	decagon
di-	Greek	twice; two	dihedral
dia-	Greek	through	diameter
dis-	Latin	apart	disjoint
dodeca-	Greek	twelve	dodecagon
e-	Latin	out of	eliminate
epi-	Greek	on	epicycloid
equi-	Latin	equal	equidistant
ex-	Latin	out of	expand
geo-	Greek	earth	geometry
hemi-	Greek	half	hemisphere
hepta-	Greek	seven	heptagon
hexa-	Greek	six	hexagon
hyper-	Greek	over	hyperbola
hypo-	Greek	under	hypocycloid
icosa-	Greek	twenty	icosahedron
in-	Latin	in	incircle
in-	Latin	not	infinite
inter-	Latin	between	interpolate

Appendix 2 Continued

Prefix	Original language	English meaning	Example
ir-	Latin	not	irrational
iso-	Greek	equal	isosceles
multi-	Latin	many	multiply
nona-	Latin	nine	nonagon
ob-	Latin	towards	oblique
octa-	Greek	eight	octagon
ortho-	Greek	straight	orthogonal
para-	Greek	beside; against	parallel
penta-	Greek	five	pentagon
per-	Latin	through; by	perspective
peri-	Greek	around	perimeter
poly-	Greek	many	polygon
pro-	Latin	forward	progression
pro-	Latin	in place of	pronumeral
quadr-	Latin	four	quadrant
re-	Latin	back; again	reflection
rect-	Latin	straight, right	rectilinear
semi-	Latin	half	semicircle
sex-	Latin	six	sexagesimal
sub-	Latin	under	submultiple
sym-	Greek	with; share	symmetry
tetra-	Greek	four	tetrahedron
trans-	Latin	across	transversal
tri-	Latin	three	triangle
uni-	Latin	one	union

Appendix 3 Mathematicians who have made a significant contribution to the development of mathematics

BC

570–504	Pythagoras	Greek	Geometry
490–430	Zeno of Elea	Greek	Logic
427–347	Plato	Greek	Geometry
384–322	Aristotle	Greek	Logic
?–300	Euclid	Greek	Geometry
287–212	Archimedes	Greek	Physics of geometry
276–192	Eratosthenes	Greek	Number and measurement
262–190	Apollonius	Greek	Geometry

AD

85–165	Ptolemy	Greek	Trigonometry
375–415	Hypatia	Greek	Algebra and conic sections
?–825	Al-Khowarizmi	Arab	Algebra
1175–1250	Leonardo Fibonacci	Italian	Algebra
1540–1603	Francois Viete	French	Algebra
1550–1617	John Napier	Scottish	Logarithms
1564–1643	Galileo Galilei	Italian	Analytic geometry
1571–1630	Johannes Kepler	German	Conics and astronomy
1596–1650	René Descartes	French	Analytic geometry
1601–1665	Pierre de Fermat	French	Analytic geometry
1623–1662	Blaise Pascal	French	Conics and probability
1643–1727	Isaac Newton	English	Calculus
1646–1716	Gottfried Wilhelm Leibniz	German	Calculus

1654–1705	Jakob Bernoulli	Swiss	Differential equations and polar coordinates
1667–1748	Johann Bernoulli	Swiss	Exponentials
1706–1749	Emilie du Châtelet	French	Calculus
1707–1783	Leonhard Euler	Swiss	Pure and applied mathematics
1718–1749	Maria Agnesi	Italian	Algebra and analysis
1776–1831	Sophie Germain	French	Number theory and analysis
1777–1855	Carl Friedrich Gauss	German	Number theory
1780–1872	Mary Somerville	English	Applied mathematics
1789–1857	Augustin-Louis Cauchy	French	Calculus
1792–1871	Charles Babbage	English	Computing
1793–1856	Nicolai Lobachevski	Russian	Geometry
1806–1871	Augustus De Morgan	English	Algebra
1815–1852	Ada Augusta Lovelace	English	Computer programming
1815–1864	George Boole	English	Algebra and logic
1820–1910	Florence Nightingale	English	Circular graphs
1850–1891	Sonya Kovalevskaya	Russian	Complex analysis
1858–1932	Giuseppi Peano	Italian	Algebra and analysis
1861–1947	Alfred North Whitehead	English	Logic
1868–1944	Grace Chisholm Young	English	Set theory and differential calculus
1872–1970	Bertrand Russell	English	Pure mathematics and logic
1879–1955	Albert Einstein	German	Relativity
1882–1935	Emmy Noether	German	Algebra
1914–1971	Hanna Neumann	Australian	Group theory